EARTH SCIENCES IN THE 21ST CENTURY

GEOSPATIAL INFORMATION AND GIS: BACKGROUND AND ISSUES

EARTH SCIENCES IN THE 21ST CENTURY

Additional books in this series can be found on Nova's website under the Series tab.

Additional E-books in this series can be found on Nova's website under the E-book tab.

EARTH SCIENCES IN THE 21ST CENTURY

GEOSPATIAL INFORMATION AND GIS: BACKGROUND AND ISSUES

SEAN C. DALLON
EDITOR

Nova Science Publishers, Inc.
New York

Copyright © 2011 by Nova Science Publishers, Inc.

All rights reserved. No part of this book may be reproduced, stored in a retrieval system or transmitted in any form or by any means: electronic, electrostatic, magnetic, tape, mechanical photocopying, recording or otherwise without the written permission of the Publisher.

For permission to use material from this book please contact us:
Telephone 631-231-7269; Fax 631-231-8175
Web Site: http://www.novapublishers.com

NOTICE TO THE READER

The Publisher has taken reasonable care in the preparation of this book, but makes no expressed or implied warranty of any kind and assumes no responsibility for any errors or omissions. No liability is assumed for incidental or consequential damages in connection with or arising out of information contained in this book. The Publisher shall not be liable for any special, consequential, or exemplary damages resulting, in whole or in part, from the readers' use of, or reliance upon, this material. Any parts of this book based on government reports are so indicated and copyright is claimed for those parts to the extent applicable to compilations of such works.

Independent verification should be sought for any data, advice or recommendations contained in this book. In addition, no responsibility is assumed by the publisher for any injury and/or damage to persons or property arising from any methods, products, instructions, ideas or otherwise contained in this publication.

This publication is designed to provide accurate and authoritative information with regard to the subject matter covered herein. It is sold with the clear understanding that the Publisher is not engaged in rendering legal or any other professional services. If legal or any other expert assistance is required, the services of a competent person should be sought. FROM A DECLARATION OF PARTICIPANTS JOINTLY ADOPTED BY A COMMITTEE OF THE AMERICAN BAR ASSOCIATION AND A COMMITTEE OF PUBLISHERS.

Additional color graphics may be available in the e-book version of this book.

LIBRARY OF CONGRESS CATALOGING-IN-PUBLICATION DATA

Geospatial information and GIS : background and issues / editor, Sean C. Dallon.
    p. cm.
  Includes index.
  ISBN 978-1-61761-432-3 (softcover)
  1. Geographic information systems. 2. Spatial analysis (Statistics) I. Dallon, Sean C.
  G70.2.G5 2010
  910.285--dc22
                                            2010028633

Published by Nova Science Publishers, Inc. † New York

CONTENTS

Preface		vii
Chapter 1	Geospatial Information and Geographic Information Systems (GIS): Current Issues and Future Challenges *Peter Folger*	1
Chapter 2	Issues Regarding a National Land Parcel Database *Peter Folger*	35
Chapter 3	The Changing Geospatial Landscape *National Geospatial Advisory Committee*	53
Chapter 4	The National Geospatial Technical Operations Center *United States Geological Survey*	79
Chapter 5	Web-Based Geospatial Tools to Address Hazard Mitigation, Natural Resource Management and other Societal Issues *United States Geological Survey*	85
Index		93

PREFACE

Geospatial information is data referenced to a place (a set of geographic coordinates) which can often be gathered, manipulated, and displayed in real time. A Geographic Information System (GIS) is a computer system capable of capturing, storing, analyzing, and displaying geographically referenced information. In recent years consumer demands have skyrocketed for geospatial information tools like GIS to manipulate and display geospatial information. The federal government and policy makers increasingly use geospatial information and tools like GIS for producing floodplain maps, conducting Census, mapping foreclosures, and responding to natural hazards such as wildfires and hurricanes. This book provides an overview on geospatial data and GIS and discusses issues that may be of interest to Congress.

Chapter 1- Geospatial information is data referenced to a place—a set of geographic coordinates—which can often be gathered, manipulated, and displayed in real time. A Geographic Information System (GIS) is a computer system capable of capturing, storing, analyzing, and displaying geographically referenced information. In recent years consumer demand has skyrocketed for geospatial information and for tools like GIS to manipulate and display geospatial information. Global Positioning System (GPS) data and their integration with digital maps has led to the popular handheld or dashboard navigation devices used daily by millions. The federal government and policy makers increasingly use geospatial information and tools like GIS for producing floodplain maps, conducting the Census, mapping foreclosures, and responding to natural hazards such as wildfires and hurricanes. For policy makers, this type of analysis can greatly assist in clarifying complex problems that may involve local, state, and federal government, and affect businesses, residential areas, and federal installations.

Chapter 2- The federal government's efforts to coordinate its geospatial activities, through the Federal Geographic Data Committee (FGDC) and the development of the National Spatial Data Infrastructure (NSDI), include a strong emphasis on land parcel data. Land parcel databases (or cadastres) describe the rights, interests, and value of property. Ownership of land parcels is an important part of the legal, financial, and real estate system of a society. The Bureau of Land Management (BLM) is assigned the role of lead agency coordinating land parcel data for federal lands, and is responsible for performing cadastral surveys on all federal and Indian lands. According to BLM, "Cadastral surveys are the foundation for all land title records in the United States and provide federal and tribal land managers with information necessary for the management of their lands."

Chapter 3- Practically overnight, access to terabytes of geographical information, much of it in three dimensions, has changed the way people work, live and play. We rely on a host of location-based technologies via our desktop computers, PDAs and even our cell phones. These services fuel a market estimated at $30 billion per year and represent a major information technology growth sector. The primary reasons mainstream commercial applications have emerged are that a wide variety of businesses have taken advantage of investments and policy decisions made by the United States government during the past thirty years, and burgeoning technology innovations. These innovations include the Internet, communications infrastructure, detailed digital mapping, robust data management systems, advancements in modeling the earth's sphere, the creation of a constellation of global positioning system (GPS) satellites, and more.

Chapter 4- The United States Geological Survey (USGS) National Geospatial Technical Operations Center (NGTOC) provides geospatial technical expertise in support of the National Geospatial Program in its development of *The National Map*, National Atlas of the United States®, and implementation of key components of the National Spatial Data Infrastructure (NSDI).

Chapter 5- Federal, State, and local government agencies in the United States face a broad range of issues on a daily basis. Among these are natural hazard mitigation, homeland security, emergency response, economic and community development, water supply, and health and safety services. The U.S. Geological Survey (USGS) helps decision makers address these issues by providing natural hazard assessments, information on energy, mineral, water and biological resources, maps, and other geospatial information.

In: Geospatial Information and GIS: Background... ISBN: 978-1-61761-432-3
Editor: Sean C. Dallon © 2011 Nova Science Publishers, Inc.

Chapter 1

GEOSPATIAL INFORMATION AND GEOGRAPHIC INFORMATION SYSTEMS (GIS): CURRENT ISSUES AND FUTURE CHALLENGES[*]

Peter Folger

SUMMARY

Geospatial information is data referenced to a place—a set of geographic coordinates—which can often be gathered, manipulated, and displayed in real time. A Geographic Information System (GIS) is a computer system capable of capturing, storing, analyzing, and displaying geographically referenced information. In recent years consumer demand has skyrocketed for geospatial information and for tools like GIS to manipulate and display geospatial information. Global Positioning System (GPS) data and their integration with digital maps has led to the popular handheld or dashboard navigation devices used daily by millions. The federal government and policy makers increasingly use geospatial information and tools like GIS for producing floodplain maps, conducting the Census, mapping foreclosures, and

---

[*] This is an edited, reformatted and augmented version of Congressional Research Service publication, Report R40625, dated June 8, 2009.

responding to natural hazards such as wildfires and hurricanes. For policy makers, this type of analysis can greatly assist in clarifying complex problems that may involve local, state, and federal government, and affect businesses, residential areas, and federal installations.

Congress has recognized the challenge of coordinating and sharing geospatial data from the local, county, and state level to the national level, and vice versa. The cost of geospatial information to the federal government has also been an ongoing concern. As much as 80% to 90% of government information has a geospatial component, according to different sources. The federal government's role has changed from being a primary provider of authoritative geospatial information to coordinating and managing geospatial data and facilitating partnerships. Challenges to coordinating how geospatial data are acquired and used—collecting duplicative data sets, for example—at the local, state, and federal levels, in collaboration with the private sector, are not yet resolved.

The federal government has recognized the need to organize and coordinate the collection and management of geospatial data since at least 1990, when the Office of Management and Budget (OMB) revised Circular A-16 to establish the Federal Geographic Data Committee (FGDC) and to promote the coordinated use, sharing, and dissemination of geospatial data nationwide. OMB Circular A-16 also called for development of a national digital spatial information resource to enable the sharing and transfer of spatial data between users and producers, linked by criteria and standards. Executive Order 12906, issued in 1994, strengthened and enhanced Circular A-16, and specified that FGDC shall coordinate development of the National Spatial Data Infrastructure (NSDI).

The high-level leadership and broad membership of the FGDC—10 cabinet-level departments and 9 other federal agencies—suggest that geospatial information is a highly regarded asset of the federal government. Questions remain, however, about how effectively the FGDC is fulfilling its mission. Has this organizational structure worked? Can the federal government account for the costs of acquiring, coordinating, and managing geospatial information? How well is the federal government coordinating with the state and local entities that have an increasing stake in geospatial information? What is the role of the private sector?

State-level geospatial entities, through the National State Geographic Information Council, also embrace the need for better coordination. However, the states are sensitive to possible federal encroachment on their prerogatives to customize NSDI to meet the needs of the states.

INTRODUCTION

The explosion of consumer demand for geospatial information and tools such as geographic information systems (GIS) to manipulate and graphically display such information has brought GIS into the daily lives of millions of Americans, whether they know it or not. Google Earth and handheld or dashboard navigation systems represent enormously popular examples of the wide variety of applications made possible through the availability of geospatial information.[1] The release of Google Earth in 2005 represented a paradigm shift in the way people understand geospatial information, according to some observers, because it offered multi-scale visualization of places and locations around the globe that was free and easy to use.[2]

Historically, the federal government has been a primary provider of authoritative geospatial information, but some argue that consumer demand for spatial information has triggered a major shift toward local government and commercial providers.[3] The federal government has shifted, with some important exceptions, to consuming rather than providing geospatial information from a variety of sources. As a result, the federal government's role has shifted as well toward coordinating and managing geospatial data and facilitating partnerships among the producers and consumers of geospatial information in government, the private sector, and academia. The challenges to coordinating how geospatial data are acquired and used—collecting duplicative data sets, for example—at the local, state, and federal levels, in collaboration with the private sector, are long-standing and not yet resolved.

In 2003 and 2004 the Subcommittee on Technology, Information Policy, Intergovernmental Relations, and the Census, part of the House Committee on Government Reform, held two hearings on the nation's geospatial information infrastructure. A common theme to both hearings was the challenge of coordinating and sharing geospatial data between the local, county, state, and national levels. Quantifying the cost of geospatial information to the federal government has also been an ongoing concern for Congress. At the hearing in 2003, Congressman Putnam stated:

> We need to understand what programs exist across the government, how much we're spending on those programs, where we're spending that money, how efficiently, or perhaps inefficiently, we share data across Federal agency boundaries, how we separate security-sensitive geospatial data from those open for public use, and how we efficiently, or perhaps inefficiently, coordinate with State and local governments and tribes.[4]

The explosion of geospatial data acquired at the local and state levels, for their own purposes and in conjunction with the private sector, underscores the long-recognized need for better coordination between the federal government and local and state authorities. At the same time, coordinating, managing, and facilitating the production and use of geospatial information from different sources, of different quality, and which was collected with specific objectives in mind has been a challenge. The federal government has recognized this challenge since at least 1990, when the Office of Management and Budget (OMB) revised Circular A-16 to establish the Federal Geographic Data Committee (FGDC) and to promote the coordinated use, sharing, and dissemination of geospatial data nationwide.[5] Executive Order 12906, issued in 1994, strengthened and enhanced the policies in Circular A-16, and specified that the FGDC shall coordinate development of the National Spatial Data Infrastructure (NSDI). Circular A-16 was itself revised in 2002, adding the Deputy Director of Management at OMB as the vice-chair of the FGDC to serve with the Secretary of the Interior.

The high-level leadership and broad membership of the FGDC—10 cabinet-level departments and 9 other federal agencies—suggest that geospatial information is a highly regarded asset of the federal government. Questions remain, however, about how effectively the FGDC is fulfilling its mission. Has this organizational structure worked? Can the federal government account for the costs of acquiring, coordinating, and managing geospatial information? How well is the federal government coordinating with the state and local entities that have an increasing stake in geospatial information? What is the role of the private sector? Congress may wish to explore these and other questions.

This report provides a primer on geospatial data and GIS and provides several examples of their use. The report discusses issues that may be of interest to Congress: sharing, coordination, and management of geospatial information, including examples of legislation. Finally, recommendations are included from several organizations for how to improve the coordination and management of geospatial information at the federal and state levels. A discussion of classified geospatial information and national security issues is beyond the scope of this report.

GIS AND GEOSPATIAL DATA: A PRIMER

GIS is a computer system capable of capturing, storing, analyzing, and displaying geographically referenced information—information attached to a location, such as latitude and longitude, or street location.[6] Geographically referenced information is also known as geospatial information. Types of geospatial information include features like highway intersections, office buildings, rivers, the path of a tornado, the San Andreas Fault, or congressional district boundaries. Information associated with a specific location is referred to in GIS parlance as an attribute,[7] such as the population of a congressional district, or amount of movement per year along the San Andreas Fault. Other terms common to geospatial data and GIS analysis are described in the box below.

The power of GIS is the ability to combine geospatial information in unique ways—by layers or themes—and extract something new. For instance, a GIS analysis might include the location of a highway intersection and the average number of vehicles that flow through the intersection throughout the day, and extract information useful for locating a business. GIS might include both the location of a river and the water depth along its course by season, and enable an analysis of the effects of development on runoff within the watershed. Overlaying the path of a severe thunderstorm with geospatial data on the types of structures encountered—homes, stores, schools, post offices—could inform an analysis of what types of building construction can survive high winds and hail.

Sources and Types of Geospatial Data

Geospatial data may be acquired by federal, state, tribal, county, and local governments, private companies, academic institutions, and nonprofit organizations. The collection and management of geospatial data are considered by many to be the costliest components of a GIS—some experts attribute close to 80% of GIS total costs to data acquisition.[8]

It should be recognized that the amount of geospatial data is expanding rapidly, the methods for acquiring geospatial data are growing, and the ways geospatial data are being used is diversifying throughout local and state governments, as well as within the federal government. It is beyond the scope of this report to encompass the universe of geospatial data and its utility to the federal government.

GEOSPATIAL AND GIS TERMINOLOGY

Attribute: descriptive information about the properties of events, features, or entities associated with a location, such as the ownership of a parcel of land, or the population of a neighborhood, or the wind speed and direction over a point on the ground.

Cadastre: the map of ownership and boundaries of land parcels.

Cartography: the study and practice of making maps.

Datum: a definition of the origin, orientation, and scale of the coordinate system and its tie to Earth.

Geocoding: assignment of alphanumeric codes or coordinates to geographically referenced data. Examples include the two-letter country codes, or the coordinates of a residence computed from its address.

Geographic Information System (GIS): a digital database in which information is stored by its spatial coordinate system, which allows for data input, storage, retrieval, management, transformation, analysis, reporting, and other activities. GIS is often envisioned as a process as much as a physical entity for data.

Geospatial data: information that identifies the geographic location and characteristics of natural and constructed features and boundaries on Earth.

Global Positioning System (GPS): a navigation system supported by a constellation of satellites placed in orbit by the U.S. Department of Defense. The satellites transmit precise microwave signals that enable GPS receivers to determine their location, speed, and direction.

Hydrography: the charting and description of bodies of water.

LIDAR: acronym for Light Detection and Ranging, a remote sensing technique that uses laser pulses to determine elevation with high accuracy, usually from an aerial survey.

Map: a two-dimensional visual portrayal of geospatial data. The map is not the data itself.

Metadata: information about the quality, content, condition, and other characteristics of data.

Orthoimagery: digital or digitized aerial photographs or images in which the pixels are geometrically rectified and geographically referenced, often including details about topography and names. The rectified orthoimage is free of geometric distortions that are part of the original photograph or image.

Polygon: a feature in GIS used to represent areas (versus a point, or a line). A polygon is defined by the lines that make up its boundary, and a point inside its boundary for identification.

However, the federal government has had and continues to have a major role in the overall framework for geospatial data, including its organization, coordination, and sharing between federal agencies and with state and local entities. The organization and coordination of geospatial data are discussed further below.

Geospatial data can be acquired using a variety of technologies. Land surveyors, census takers, aerial photographers, police, and even average citizens with a GPS-enabled cell phone can collect geospatial data using GPS or street addresses that can be entered into GIS.[9] The attributes of the collected data, such as land-use information, demographics, landscape features, or crime scene observations, can be entered manually or, in the case of a land survey map, digitized from a map format to a digital format by electronic scanning. Remote sensing data from satellites is acquired digitally and communicated to central facilities for processing and analysis in GIS. Digital satellite images, for example, can be analyzed in GIS to produce maps of land cover and land use. When different types of geospatial data are combined in GIS (e.g., through combining satellite remote sensing land use information with aerial photograph data on housing development growth), the data must be transformed so they fit the same coordinates. GIS uses the processing power of a computer, together with geographic mapping techniques (cartography), to transform data from different sources onto one projection[10] and one scale so that the data can be analyzed together.

Geospatial Data from Local, State, and Federal Governments and the Private Sector

Local and state governments provide geospatial data for use in GIS for a variety of public services such as land records, property taxation, local planning, subdivision control and zoning, and others.[11] Some observers note that local governments often contract with private sector companies to acquire more recent and higher-resolution data than what is available to the federal government.[12] Whether and how the most up-to-date and detailed geospatial information is made available to users other than the local government for whom the data were acquired are longstanding issues. For example, in the immediate aftermath of a natural disaster, such as Hurricane Katrina in 2005, it may be important for the federal government to acquire the most current and detailed geospatial information about the disaster area. In many instances, however, impediments to data sharing such as lack of interoperability between systems, restrictions on use, concerns about data security, and a lack of knowledge about what data exist and where the data can be found could hinder a timely and effective emergency response.[13]

The federal government sometimes acquires or contracts to acquire geospatial data for federal needs, such as for updating floodplain maps from paper flood insurance rate maps to a digital format. Assessing and updating floodplain maps on a periodic basis is required by law,[14] and the Federal Emergency Management Agency (FEMA) has spent over $1.4 billion since FY2003 to convert paper flood insurance rate maps (FIRMs) to digital flood insurance rate maps (DFIRMs) and to produce a format usable in GIS.[15] Simply converting paper maps to digital formats does not necessarily improve their accuracy, which often depends on the resolution of the original data. New techniques for collecting more data, such as Light Detection and Ranging (LIDAR), will help produce more accurate floodplain maps.

Example: Using LIDAR for Floodplain Mapping

There is no single nationwide elevation dataset of sufficient resolution and accuracy to make floodplain maps that meet FEMA requirements. A fundamental requirement for accurate flood maps is accurate elevation data, which are used to draw the boundaries for the 1-in-100 chance annual flood hazard (sometimes referred to as the 100-year flood). The USGS National Elevation Dataset is a primary data source that FEMA uses to produce flood maps, but it has a level of uncertainty about 10 times larger than FEMA defines as acceptable for floodplain mapping.[16] The USGS National Elevation Dataset includes some high resolution, more accurate elevation data, but most of the USGS dataset is of poorer resolution. Alternate sources of more accurate elevation data exist, but are not available nationwide. One of these sources is provided using LIDAR,[17] which can be used to collect high resolution elevation data.[18] Because of this data gap, a National Research Council report recommends that FEMA should increase its collaboration with federal, state, and local government agencies to acquire high resolution and accurate elevation data across the nation.[19]

Geospatial data are increasingly acquired and provided by the private sector, and many companies as well as professional organizations support and promote the role of private sector data providers. One organization, the Management Association for Private Photogrammetric Surveyors (MAPPS), bills itself as the only national association exclusively composed of private geospatial firms.[20] MAPPS itself is a member of a larger coalition—the

Coalition of Geospatial Organizations (COGO). COGO is comprised of 15 geospatial-related organizations and associations.[21]

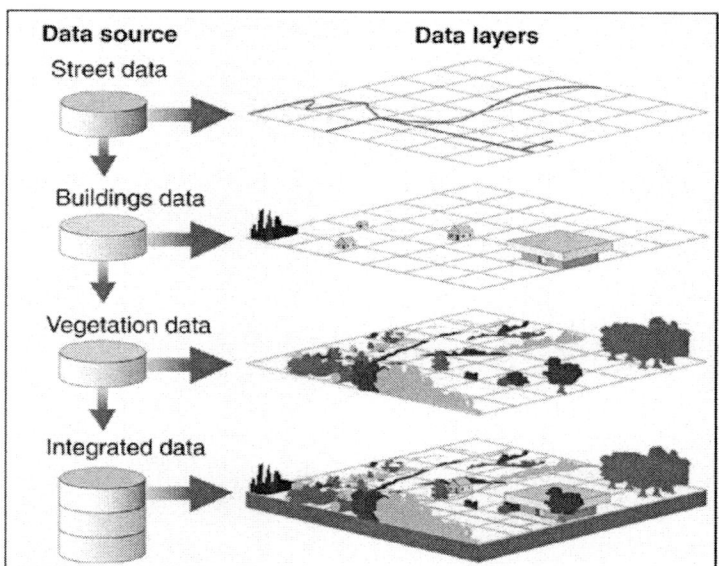

Source: GAO (2004), p. 5.

Figure 1. Example of GIS Data Layers or Themes

GIS Layers or Themes

The attributes of different types of geospatial data—such as land ownership, roads and bridges, buildings, lakes and rivers, counties, or congressional districts—can each constitute a layer or theme in GIS. (See **Figure 1** for a schematic representation of data layers in GIS.) GIS has the ability to link and integrate information from several different data layers or themes over the same geographic coordinates, which is very difficult to do with any other means. For example, GIS could copmbine a major road from one data layer as the boundary dividing land zoned for commercial development with the location of wetlands from another data layer. Precipitation data, from a third data layer, could be combined with a fourth data layer that shows streams and rivers. GIS could then be used to calculate where and how much runoff might flow from the commercial development into the wetlands. Thus the power of GIS analysis can be used to create a new way to interpret information that would otherwise be very difficult to visualize and analyze.

EXAMPLES OF WHY AND HOW GEOSPATIAL INFORMATION IS USED

California Wildfires

Geospatial information is data referenced to a place—a set of geographic coordinates—which can often be gathered, manipulated, and displayed in real time. Timeliness is an important factor for some uses of geospatial information. An example is the southern California wildfires during 2008. One of the worst fires in the region, known as the Sylmar fire, sparked on the evening of November 14, 2008, and swept quickly through 11,000 acres, destroying more than 600 structures in the Sylmar section of Los Angeles before it was contained.[22] The fire forced the evacuation of thousands of residents from their homes. The speed of the fire's progress made it difficult to know where the fire was heading and to visualize escape routes. In addition, the fire jumped Interstate 210 (I-210) and Interstate 5 (I-5), two major routes of egress, on Saturday, November 15.

To assist in real-time decision making, the fire's progress was posted on the Internet in near realtime by several organizations, using reports from the ground, and the information about the fire was displayed on underlying street maps (showing where the fire crossed I-5 and I-210), terrain maps, and satellite images. (See **Figure 2**.) The Sylmar fire example underscores the informational power available when geospatial information is combined with tools for displaying the information, such as GIS and the Internet. In this instance, timeliness—the ability to post the geospatial information quickly—enhanced its value to the data users, citizens trying to avoid the path of the fire.

Although timeliness is often important, the analytical power resulting from combining geospatial information with GIS more typically underscores its value to policy makers at all levels. GIS often provides for unique analyses of disparate types of information—linked by their spatial coordinates—to help resolve policy questions. For policy makers, this type of analysis can greatly assist in clarifying complex problems that may involve local, state, and federal government, and may affect businesses, residential areas, and federal installations.

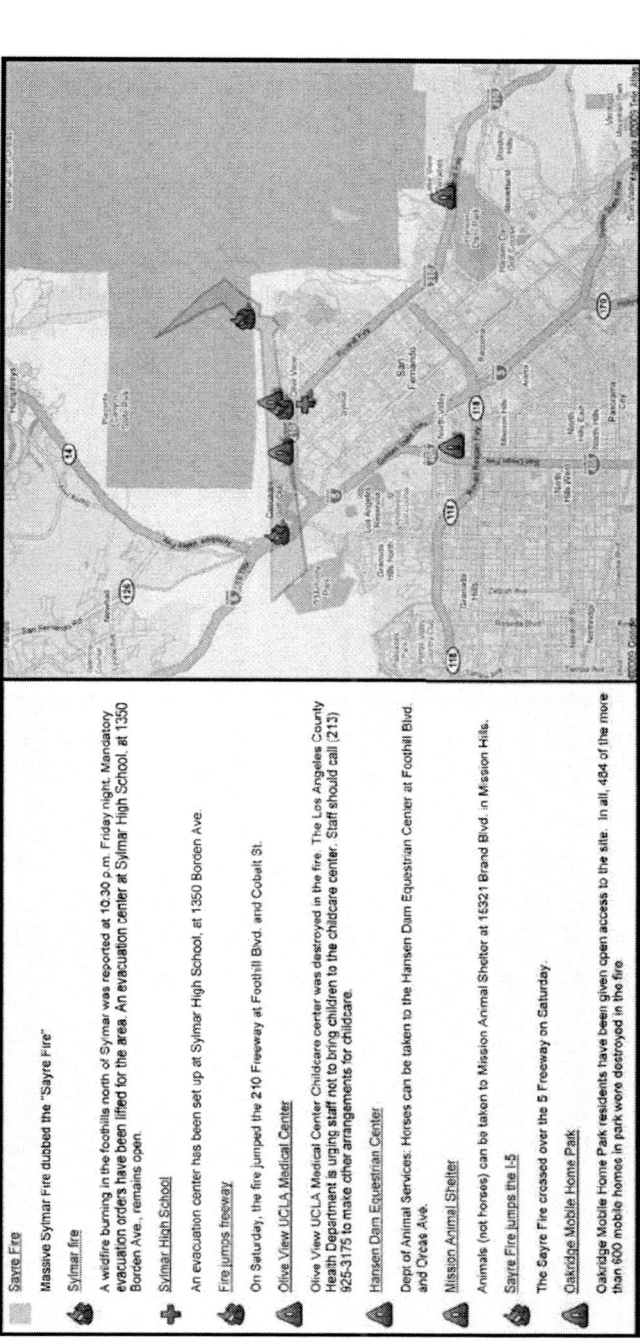

Source: ABC Eyewitness News, Sylmar Wildfire, created Nov. 15, 2008, updated Nov. 20, 2008. See http://maps.google=1008669070826291704785e21707089258&t=h&source=embed&ll=34.314638,-118.436834&spn=0.251809,0.528717&z=12. Modified by CRS.

Notes: the path of the fire with the annotation is shown with an underlying street map. The original interactive website also allows the user to choose an underlying terrain map or satellite image map.

Figure 2. Snapshot of the Path of the 2008 Sylmar Fire Near Los Angeles, CA

Base Realignment and Closure (BRAC) Program

The Base Realignment and Closure (BRAC) program is the process by which excess military facilities are identified and transferred to other federal agencies or disposed of. The City of Virginia Beach, VA, used GIS in its response to the 2005 BRAC Commission's recommendation to realign Naval Air Station Oceana, located near the population center of the city. The BRAC Commission was concerned that the city's land use was encroaching on the air station; in particular, the city was impinging on the noise zones and accident potential zone (APZ) around the air station. Because the recommended realignment of Oceana would likely cause Virginia Beach to suffer significant economic losses, the city sought to establish a baseline—using GIS—to understand the status of encroachment. In addition, the GIS analysis could inform city leaders about how to modify the municipal land use ordinance to prevent encroachment on the air station and forestall its realignment.

To establish a baseline, city planners and GIS analysts overlaid noise zones and APZ, property, land use, zoning, and other sets of geospatial data—known as attributed boundary layers—to determine current land use and development. Within the GIS analysis, these sets of geospatial data were joined with land parcel information, and with various external databases held within the planning, real estate assessor's, and commissioner's offices. By combining geospatial data with non-spatial data, the GIS analysts helped land planners determine how the land around the air station was being used, and therefore its compatibility with the Navy's requirements. (See **Figure 3**.) The GIS analysis also enabled the city to summarize property values and acreage by its use: undeveloped, commercial, or residential.

GIS helped the Virginia Beach city planners to identify on one map all of the land use around the air station (**Figure 3**). GIS analysts also provided a model of underdeveloped land—land that had additional existing by-right development capacity—but which if developed could exacerbate the encroachment problem for the Navy. As a result of the GIS analysis, city planners recommended a change to the municipal land ordinance to prevent potential future incompatible development. Naval Air Station Oceana has not been relocated from Virginia Beach.

Geospatial Information and Geographic Information Systems... 13

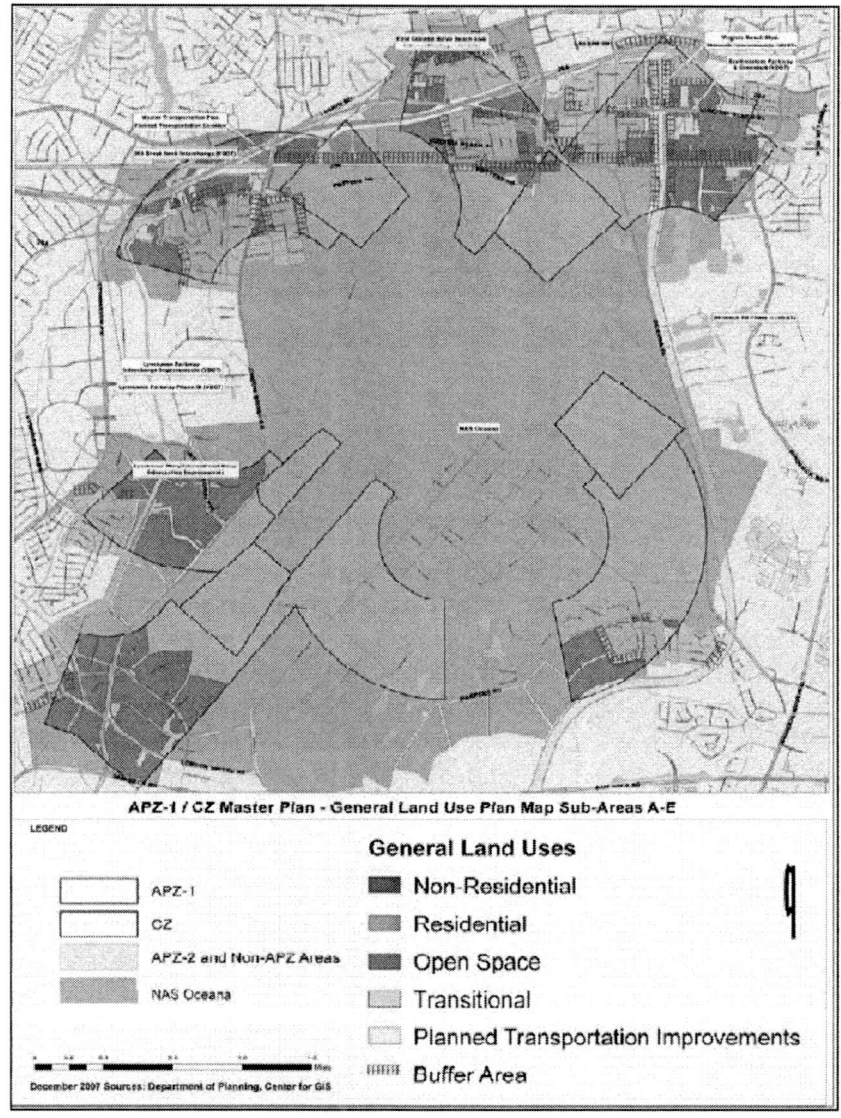

Source: Figure 5, page 19, APZ-1/Clear Zones Master Plan, Second Progress Report, N.A.S. Oceana Encroachment, City of Virginia Beach.
Notes: APZ is Accident Potential Zone; CZ is Clear Zone.

Figure 3. GIS Analysis of Naval Air Station Oceana, Virginia Beach, VA (example showing city land use encroachment)

14   Peter Folger

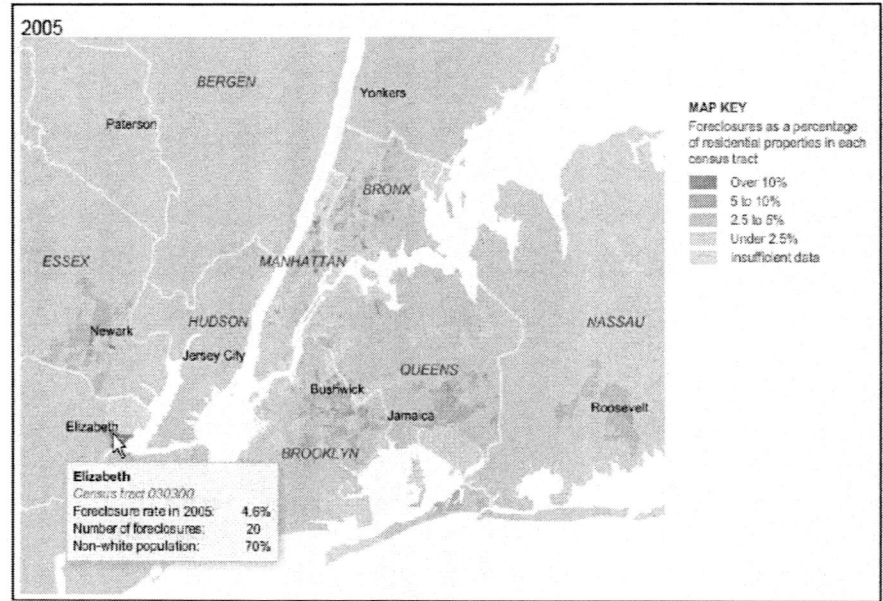

Source: New York Times, May 15, 2009, at http://www.nytimes.com/interactive/
2009/05/ 15/nyregion /0515-foreclose.html. Modified by CRS.
Notes: The online interactive version allows the reader to point and click on any
Census tract in the region. Census tract 030300 is shown here for illustration
purposes.

Figure 4. Snapshot of Interactive Map Showing Foreclosure Percentage by Census
Tract in the New York City Area, 2005

Mapping Foreclosures

On May 15, 2009, the *New York Times* published an online interactive map showing foreclosures as a percentage of residential properties in each Census tract in the New York City region.[23] The map shows Census tracts coded by color to represent the foreclosure rate, and as the cursor is moved over each Census tract, the map shows a pop-up window disclosing the foreclosure percentage, the number of foreclosed residences, and the percentage on the non-white population in each tract. In addition, the map allows the reader to compare foreclosure rates for each year since 2005. **Figures 4** and **5** are snapshots from the map for the years 2005 and 2008.

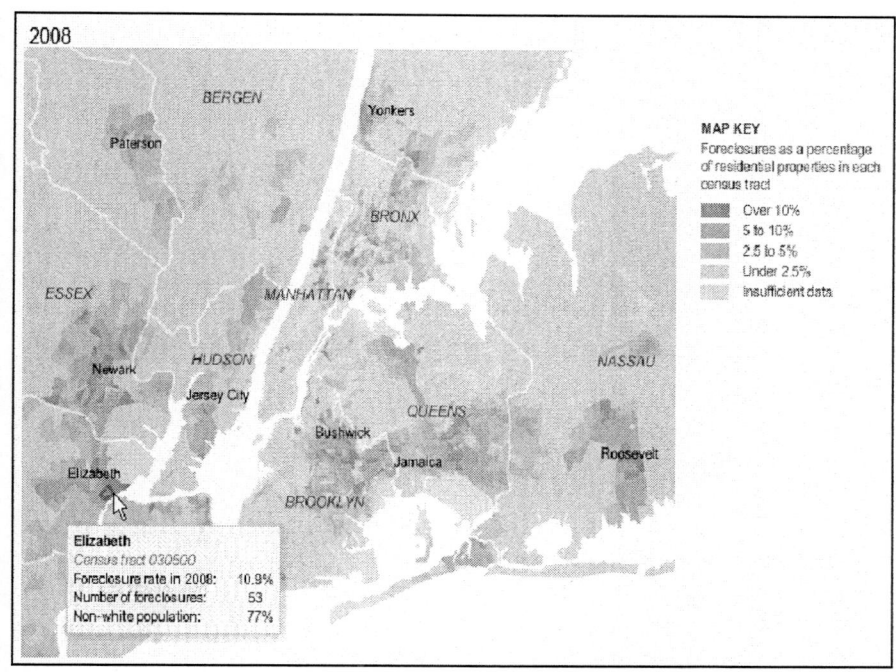

Source: New York Times, May 15, 2009, at http://www.nytimes.com/interactive/ 2009/ 05/15/nyregion/0515-foreclose.html. Modified by CRS.

Notes: the online interactive version allows the reader to point and click on any Census tract in the region. Census tract 030500 is shown here for illustration purposes.

Figure 5. Snapshot of Interactive Map Showing Foreclosure Percentage by Census Tract in the New York City Area, 2008

By using the zoom tool provided with the map, the reader can zoom in on specific residential properties, represented by red dots, along with street names. This type of visualization, combining detailed geospatial information with demographic and financial data, lends itself to further analysis such as understanding foreclosure patterns and whether proximity to foreclosed properties has an effect on property values. Some researchers call this the "contagion effect" of foreclosed properties. One report documented how this effect discounted property values as a function of distance from foreclosed homes, and showed that the discount effect dropped off sharply with distance.[24] This type of spatial analysis of foreclosure effects, with the visualization provided by GIS maps such as the *New York Times* example, can help inform policy makers about the nature of foreclosure patterns.

ISSUES WITH ORGANIZATION AND MANAGEMENT, DATA SHARING, AND COORDINATION

Producing floodplain maps, conducting the Census, planning ecosystem restoration, and assessing vulnerability and responding to natural hazards such as hurricanes and earthquakes are examples of how federal agencies use GIS and geospatial information to meet national needs. The amount of government information that has a geospatial component—such as address or other reference to a physical location—is as much as 80%, according to the Department of the Interior.[25] According to one report, geospatial-related industries generate at least $30 billion annually,[26] and the U.S. Bureau of Labor cites statistics that suggest the geospatial sector has been growing by about 35% per year, with the commercial side growing at 100% per year.[27]

Overarching challenges are:

- the organization and management of the vast array of geospatial information that is acquired at many levels and that has a variety of potential uses;
- data sharing, particularly among local, state, and federal stakeholders, each of whom may have a need for the same or similar data; and
- coordination among federal agencies and with other stakeholders, such as the administration and management by different agencies of all the federal lands in the United States.

Organization and Management of Geospatial Data

The need to organize and manage geospatial data between federal agencies and between the federal government, local and state authorities, the private sector, and academia is a recurring theme. It recurs, in part, because it is widely recognized that collecting data multiple times for the same purpose is wasteful and inefficient, yet it continues to occur. Alternatively, geospatial data collected to meet the requirements of, for example, a local government, could be made useful to the state or federal government if the data meet a set of basic and consistent guidelines and protocols. In fact, organizational structures exist at the federal and state levels to identify and promulgate the efficient sharing, transfer, and use of geospatial information. Ideally, these efforts would produce a national spatial data infrastructure, or NSDI. Some members of the geospatial community have indicated that the past efforts to create a national spatial data infrastructure have not met expectations, and have recently called for a new effort to build a "national GIS" or a "NSDI 2.0." (See discussion below on these proposals.) In addition to promoting the efficiency and interoperability of such a national system, some promote NSDI as "digital infrastructure" on par with other parts of the nation's critical infrastructure—such as roads, pipelines, telecommunications—and underscore its role in the national economy and in national security.

Background

The federal government has recognized the need to organize and coordinate the collection and management of geospatial data since at least 1990, when OMB revised Circular A-16 to establish the FGDC, and to promote the coordinated use, sharing, and dissemination of geospatial data nationwide.[28] OMB Circular A-16 also called for development of a national digital spatial information resource to enable the sharing and transfer of spatial data between users and producers, linked by criteria and standards. President Clinton issued Executive Order 12906 on April 11, 1994, to strengthen and enhance the general policies described in Circular A-16, and to specify that the FGDC shall coordinate development of the National Spatial Data Infrastructure (NSDI). OMB revised Circular A-16 most recently on August 19, 2002, to affirm the NSDI as "the technology, policies, standards, human resources, and related activities necessary to acquire, process, distribute, use, maintain, and preserve spatial data." The revised circular incorporated Executive Order 12906, and added the Deputy Director of Management at OMB as the vice-chair of the FGDC to serve with the Secretary of the Interior.

In one sense, the FGDC exists to foster development and implementation of the NSDI. The NSDI includes the processes and relationships that facilitate data sharing across all levels of government, academia, and the private sector. Ultimately, the NSDI is intended to be the base resource and structure among geospatial data providers and users at the national, state, and local level.

The Federal Geographic Data Committee (FGDC)

Under the revised Circular A-16, 19 members comprise the FGDC (see **Table 1**). The U.S. Geological Survey, Department of the Interior, provides administrative support through the FGDC Secretariat.[29] According to Circular A-16, all federal agencies responsible for geospatial data themes are required to be members of the FGDC. Further, Circular A-16 directs the FGDC to lead and support the NSDI strategy, spatial data policy development, management, and operational decision making. As the overall coordinating entity for the NSDI, the FGDC has broad responsibilities that include all spatial data and geographic information systems activities that are financed directly or indirectly, in part or in whole, by federal funds.

The National Spatial Data Infrastructure (NSDI)

The FGDC facilitates the NSDI in cooperation with organizations from state, local, and tribal governments, the academic community, and the private sector. As specified in Circular A-16, cooperation is necessary to realize the overall vision of the NSDI, which is to assure that spatial data from multiple sources—not just federal sources—are available and easily integrated to enhance the understanding of our physical and cultural world. The five components of the NSDI are:

- *Data themes*. These are geodetic control, orthoimagery, elevation and bathymetry, transportation, hydrography, cadastre, and governmental units.
- *Metadata*. These are information about the data, its content, source, accuracy, method of collection, and other descriptions that help ensure the data are used appropriately. OMB Circular A-16 specifies that all spatial data collected or derived directly or indirectly using federal funds will have FGDC metadata.

Table 1. Members of the Federal Geographic Data Committee (FGDC)

Dept. of Agriculture	Environmental Protection Agency
Dept. of Commerce	Federal Emergency Management Agency
Dept. of Defense	General Services Administration
Dept. of Energy	Library of Congress
Dept. of Health and Human Services	National Aeronautics and Space Administration
Dept. of Housing and Urban Development	National Archives and Records Administration
Dept. of the Interior	National Science Foundation
Dept. of Justice	Tennessee Valley Authority
Dept. of State	Office of Management and Budget
Dept. of Transportation	

- *National Spatial Data Clearinghouse.* The Clearinghouse is an electronic service providing access to documented spatial data and metadata from distributed data sources. The Clearinghouse is intended to provide access to NSDI for spatial data users.
- *Standards.* These are common and repeated rules, conditions, guidelines or characteristics for data, and related processes, technology, and organization. OMB Circular A-16 specifies that international standards and protocols must be used for NSDI, to broaden the global use of federal data and services. The FGDC is responsible for developing and promulgating the standards after receiving broad input from data users and providers.
- *Partnerships.* OMB Circular A-16 directs federal agencies to promote and fully utilize partnerships that promote cost-effective data collection, documentation, maintenance, distribution, and preservation strategies that leverage the federal resources. In addition to federal, state, and tribal governments, these partnerships are supposed to include private-sector geographic, statistical, demographic, and other business information providers and users.

Other Activities and Components of FGDC and NSDI

Geospatial One-Stop

According to the FGDC 2007 Annual Report,[30] the Geospatial One-Stop portal is the official means of accessing metadata resources, which are published through the National Spatial Data Clearinghouse and which are managed in NSDI. Geospatial One-Stop focuses on the discovery and access of geospatial information.[31] The Geospatial One-Stop is described as one of the three national geospatial initiatives that share the goal of building the NSDI along with NGDC itself and The National Map (described below). The FGDC focuses on policy, standards, and advocacy, and The National Map focuses on integrated, certified, base mapping content.

The National Map

According to the USGS, The National Map is envisioned as a consistent framework for geographic knowledge nationwide, and will be available as an online and interactive map service.[32] The National Map is the future product that would supplant the paper versions of topographic maps that the USGS has produced for decades. It would allow users to combine geographic information from other sources with the USGS topographic foundation data. The National Map would provide information such as high-resolution digital imagery from satellites and aerial photographs, high-resolution surface elevation data, land cover data, geographic names, and other features. Currently, The National Map is in an initial stage that can provide nationwide coverage at limited resolutions for transportation, hydrography, elevation, land cover, and cultural features. According to the USGS, The National Map will capture and integrate data in a process of continuous update, rather than by regularly scheduled cycles of review and revision. The National Map will face challenges, however, in integrating data from a variety of sources, perhaps at different scales and different resolutions, and in managing inconsistent or incomplete metadata.

USGS Geospatial Liaison Network

The National Geospatial Program (NGP) at the USGS includes the USGS Geospatial Liaison Network, which consists of USGS employees who serve as liaisons in NSDI partnership offices across the country.[33] The liaisons are intended to represent and coordinate NGP initiatives in state and local agencies, in addition to other federal agencies, in support of NSDI, The National Map, and Geospatial One-Stop. The liaisons work with statewide coordinating councils and seek partnerships with not-for-profit organizations, the private sector, universities, and consortia to support the goals of NSDI. According to the USGS, each liaison is the "local face" of the USGS NSDI and NGP.

Each state is assigned a liaison under the network, although some liaisons may cover more than one state.[34] The liaisons commonly work with the formal state GIS coordinators or councils, or with other individuals or regional groups in states where a formal GIS or geospatial coordinator or council does not exist. A large portion of the liaison's efforts is devoted to coordinating with state-level and other stakeholders on geospatial data acquisition, although some of the focus has been shifting to maintaining geospatial data as well as acquiring it.[35] In states with large federal land holdings, such as some western states, geospatial liaisons may devote relatively more time to coordinating with federal land management agencies such as the USFS, National Park Service, or BLM.[36]

Why Data Sharing between Local, State, and National Levels Is Important

The National Research Council (NRC) has reported that the value of geospatial data is better accepted at the county level than it was in the past, especially land parcel, or cadastral, data. The benefits of sharing geospatial data so that what is produced locally can by used for national needs, however, is not as widely acknowledged.[37] In the case of land parcel data specifically, many local governments create data for their own use and do not see how a national effort would bring local benefits. The NRC notes, however, that the need for complete national land parcel data has become urgent particularly for at least one application—emergency response. During the Hurricane Katrina disaster, some critical land parcel data that was needed by emergency responders, public officials, and even insurance companies was not readily available or did not exist.[38] Further, the NRC report asserts that many of the property fraud cases associated with the hurricanes of 2005 were the direct result of poor or nonexistent geospatial data, specifically land parcel data.[39]

Example: Floodplain Mapping

In one sense, floodplain mapping represents the archetypical example of how GIS and geospatial data can be shared to fulfill national and local needs. In this case, the shared need is for accurate floodplain maps. Floodplain mapping also exemplifies the challenge to collecting and sharing geospatial data of sufficient accuracy to meet the needs of local, state, and federal data users and authorities.

In a recently released report, the NRC observed that high quality, digital mapping of floodplains using the most accurate elevation data is essential to communicate flood hazards, set flood insurance rates, and regulate development in flood-prone areas.[40] Between 2003 and 2008, the FEMA invested approximately $1 billion in the Map Modernization Program, a large-scale effort to collect new elevation data, update existing data, and digitize older paper flood maps.[41] State governments and local partners also contributed considerable funding to the effort. The FEMA effort produced digital flood maps covering 92% of the nation's population; however, only 21% of the population has flood maps that fully meet FEMA's own data quality standards. As a result, insurance companies, lenders, realtors, and property owners who depend on the flood maps to determine flood insurance needs, plan for development, and prepare for floods still have to deal with uncertainties inherent in the less accurate flood maps.

Challenges to Coordinating

Several efforts to coordinate geospatial data among federal agencies have proven difficult to achieve. The National Map is an example of a work-in-progress attempting to integrate data from a variety of sources and produce a product that is widely available and useful to many users. In an example cited by the GAO in its 2003 testimony, the U.S. Forest Service (USFS) tried to create a national-level GIS for the forest ecosystem, but had to reconcile data from a variety of incompatible locally developed systems, which used a variety of standards for each forest and district. Most of the USFS effort went into reconciling the different data sets. Ultimately the USFS had to adopt the lowest-resolution format to maintain full coverage of all the forests, and could not use the higher-resolution local data.[42]

The National Integrated Land System (NILS) is another example of an ongoing effort to coordinate and integrate federal land data among several agencies. The Bureau of Land Management (BLM) is the designated custodian for federal land parcel information and ownership status.[43] The federal government owns approximately 650 million acres, about 29% of all land in the United States. Three federal agencies in addition to the BLM administer most federal lands: the USFS, Fish and Wildlife Service, and the National Park Service.[44] In an effort to develop a single representation of federal lands, the BLM and USFS launched a joint project called the National Integrated Land System (NILS), billed as a partnership between the federal agencies and states, counties, and private industry to provide a single solution to managing federal land parcel information in a GIS environment.[45] A limited amount of federal land data is available through NILS, which is currently in a project or prototype phase, and the project makes current information and tools available through its GeoCommunicator component.[46]

Both NILS and The National Map represent federal efforts to foster interagency sharing of data into a single product providing national coverage of federal land holdings and topography respectively. The utility of both efforts is limited by the quality, accuracy, and completeness of the underlying geospatial data. The National Map, as currently envisioned, will provide topographic information at the 1:24,000 scale, meaning that roughly one inch on the map equals 2,000 feet. That scale will likely limit The National Map's usefulness for depicting, for example, floodplain boundaries that meet the requirements for FEMA floodplain maps. Also, at some point in the future NILS presumably could provide one-stop shopping for an accurate assessment of the amount of federal land currently administered by each land management agency in the Department of the Interior and for the USFS. Currently, however, the best method for obtaining an accurate tally of federal lands is to contact each land management agency directly and request their most up-to-date data in tabular form.[47] Legislation has been introduced in Congress to address some of these challenges.

The Federal Land Asset Inventory Reform Act of 2009

On March 16, 2009, Representative Kind introduced H.R. 1520, the *Federal Land Asset Inventory Reform Act of 2009*, which would require the Secretary of the Interior to develop a multipurpose cadastre of federal "real property." The legislation defines cadastre as an inventory, and defines federal "real property" as land, buildings, crops, forests, or other resources still attached to or within the land or improvements or fixtures permanently attached to the land or structures on it. The bill requires the Secretary to coordinate with the FGDC pursuant to OMB Circular A-16, to integrate the activities under the legislation with similar cadastral activities of state and local governments, and to participate in establishing standards and protocols that are necessary to ensure interoperability of the geospatial information of the cadastre for all users. Similar legislation was introduced in the Senate and House in the 110th Congress.[48]

By developing the cadastre, the legislation is intended to improve federal land management, resource conservation, environmental protection, and the use of federal real property. As noted above, the BLM currently has responsibility for maintaining federal land parcel information and ownership status, and it is not clear if H.R. 1520 would expand the current geospatial activities at BLM, shift the custodial responsibilities to another agency, or result in a different approach or program. Some supporters of the bills introduced in the 110th Congress indicated that existing inventories of federal real property are old, outdated, and inaccurate.[49] Observers also note that the federal government lacks one central inventory that coordinates all the inventories into one usable database.[50]

The Ocean and Coastal Mapping Integration Act

The Ocean and Coastal Mapping Integration Act, introduced as S. 174 and H.R. 365 in the 111th Congress, was enacted into law as Subtitle B of Title XII of the Omnibus Public Land Management Act of 2009 (P.L. 111-11). The act establishes a federal program to develop a coordinated and comprehensive mapping plan for the coastal waters including the exclusive economic zone and continental shelf, and the Great Lakes. In establishing the program, the act addresses issues of data sharing and cost-effectiveness by fostering cooperative mapping efforts, developing appropriate data standards, and facilitating the interoperability of data systems. Further, the program established under the act would develop these standards to be consistent with the requirements of the FGDC, so that the data collected in support of mapping are useful not only to federal government, but also to coastal states and other entities. The theme of coordinating activities is underscored in several places in the act, specifically with other federal efforts such as the Digital Coast,[51] Geospatial One-Stop (discussed above), and the FGDC, as well as international mapping activities, coastal state activities, user groups, and nongovernmental entities.

The challenge to collect and manage the geospatial data needed to meet the requirements of the act is daunting, given the array of federal agencies, affected states, local communities, businesses, and other stakeholders who have an interest in coastal mapping. Moreover, the stakeholders require wide and disparate types of data—such as living and nonliving coastal and marine resources, coastal ecosystems, sensitive habitats, submerged cultural resources, undersea cables, aquaculture projects, offshore energy projects, and others. A coordinated effort is more likely to produce a robust coastal mapping effort called for in the act. Congress could view the development of the ocean and coastal mapping plan and its implementation as a test case: how to manage a large data collection effort—cost-effectively and cooperatively—that reaches across all levels of government and includes interest groups, businesses, NGOs, and even international partners.

Non-Federal Stakeholders

National Geospatial Advisory Committee

A National Geospatial Advisory Committee (NGAC) was formed in early 2008 to provide advice and recommendations to the FGDC on management of federal geospatial programs, development of the NSDI, and implementation of the OMB Circular A-16. The Secretary of the Interior named 28 individuals to the committee on January 28, 2008; these members represent the private sector, nonprofits, academia, and governmental agencies.[52] As part of its charter, NGAC provides a forum to convey views representative of non-federal stakeholders in the geospatial community.

In its January 2009 report, *The Changing Geospatial Landscape*, NGAC noted that as geospatial data production has shifted from the federal government to the private sector and state and local governments, new partnerships for data sharing and coordination are needed. Specifically:

> the hodgepodge of existing data sharing agreements are stifling productivity and are a serious impediment to use even in times of emergency.... When the federal government was the primary data provider, regulations required data to be placed in the public domain. This policy jump-started a new marketplace and led to the adoptions of GIS capabilities across public and commercial sectors. However, these arrangements are very different when data assets are controlled by private companies or local governments.[53]

NGAC observed further that the federal government's need for land parcel (cadastral) data, which is also emphasized by the National Research Council, is missing an arrangement for acquiring the detailed property-related data necessary to make decisions during times of emergency. The report suggests that detailed land parcel data—its use, value, and ownership—is needed by FEMA, the USFS, and the U.S. Department of Housing and Urban Development for emergency preparedness, response to hurricanes or wildfires, or to monitor the current foreclosure problems.[54]

NGAC Recommendations to the New Administration

In October 2008 NGAC sent recommendations to the 2008-2009 Presidential Transition Team for improving the federal role in coordinating geospatial activities, for making changes to the *U.S. Code* pertaining to non-sensitive address data, and for enhancing geospatial workforce education.[55] Most recommendations pertained to how the federal government could better coordinate geospatial partnerships with state, local, and tribal governments, the private sector, and the academic community, such as recommendations to:

- establish a geospatial leadership and coordination function immediately within the Executive Office of the President; the geospatial coordination function should be included in the reauthorization of the E-Government Act;
- require OMB and FGDC to strengthen their enforcement of OMB Circular A-16 and EO 12906;
- establish/designate Geographic Information Officers with each department or agency with responsibilities stipulated within OMB Circular A-16;

- establish and oversee an Urgent Path[56] forward for implementation of geospatial programs necessary to support current national priorities and essential government services underpinning the NSDI; and
- continue NGAC.

Access to Geospatial Information

In its recommendations, NGAC also calls for revising "restrictive statutory language as it pertains to non-sensitive address data in Title 13 U.S. Code and to 'geospatial' data in Section 1619 of the 2008 Farm Bill." In Title 13 Congress delegates responsibility for conducting the Census to the Secretary of Commerce. The law contains provisions for not disclosing or publishing private information that identifies an individual or business (Sections 9 and 214 of Title 13). The Census Bureau is forbidden to publish any private information—such as names, addresses, telephone numbers—that identifies an individual or business.[57] Interestingly, this type of geospatial information is available for some localities[58] in the United States; however, it is not provided by the Census Bureau. A proposal to amend portions of Title 13 and make geospatial data collected by the Census Bureau more accessible will likely raise issues about the privacy of personal data collected by the federal government; the value of such data for emergency management; disaster preparation; other local, regional, and national needs; and the various tradeoffs between privacy concerns and the accessibility to geospatial data.

Section 1619 of the 2008 farm bill (P.L. 110-246) prohibits disclosure of geospatial information about agricultural land or operations, when the information is provided by an agricultural producer or owner of agricultural land and maintained by the Secretary of Agriculture. Certain exceptions, contained in Section 1619 of the 2008 farm bill, apply to the prohibition. NGAC has taken the position that the statutory language could be revised to enhance the value of the geospatial data while not compromising privacy.[59]

National States Geographic Information Council (NSGIC)

At the national level, the FGDC exists to promote the coordinated development, use, sharing, and dissemination of geospatial data. At the state level, NSGIC exists to promote the coordination of statewide geospatial activities in all states, and to advocate for the states in national geospatial policy initiatives to help enable the NSDI.[60] NSGIC ties its activities to the NSDI by promoting the development of Statewide Spatial Data Infrastructures

(SSDI), under a partnership called the 50-States Initiative, which ideally would lead to the creation of an SSDI for each state. In this vision, each state's SSDI would enable coordination between geospatial data producers and consumers at all levels within the state, and allow the state to share geospatial data with the national geospatial structure envisioned as the NSDI. The emphasis on organization and coordination of geospatial data and activities is seen as critical to reducing costs to states and the federal government by eliminating data redundancy—collecting the data once, using it many times—and by setting standards that allow different users to share geospatial data regardless of who collects it.

NSGIC identified 10 criteria that define a "model" state program necessary to develop effectively coordinated statewide GIS activities, and thus reduce inefficiency and waste. These include:

1. strategic and business plans;
2. a full-time, paid, GIS coordinator and staff;
3. clearly defined authority and responsibility for coordination;
4. a relationship with the state chief information officer;
5. a political or executive champion for coordinating GIS;
6. a tie to the national spatial data infrastructure and clearinghouse programs;
7. the ability to work with local governments, academia, and the private sector;
8. sustainable funding, especially for producing geospatial data;
9. the authority for the GIS coordinator to enter into contracts; and
10. the federal government working through the statewide coordinating body.

Not all states have fully embraced the need for statewide coordination of GIS activities, and states differ in their structure and organization of geospatial data among and between state, county, and local entities. For example, some states such as Arkansas share geospatial data across agencies in a very open manner; other states such as New York require more formal agreements or have restrictions to sharing data that include critical infrastructure. (Nonetheless, some level of data sharing does occur, even in the more restrictive states.[61])

Imagery for the Nation

A priority for NSGIC is a program under development, called Imagery for the Nation (IFTN), that would collect and disseminate aerial and satellite imagery in the form of digital orthoimagery. In its description of the program, NISGIC notes that digital orthoimagery is the foundation for most public and private GIS endeavors. Further, NSGIC states that as many as 1,300 different government entities across the nation are developing digital orthoimagery products, "leading to higher costs, varying quality, duplication of effort, and a patchwork of products."[62] IFTN represents an effort to establish one coherent set of geospatial data—arguably one of the most important layers in a GIS, orthoimagery—that is organized for the benefit of many stakeholders at the federal, tribal, regional, state, and local levels.

As proposed, IFTN would involve two programs: (1) the existing National Agricultural Imagery Program (NAIP) administered by the U.S. Department of Agriculture, and (2) a companion program administered by the USGS. The NAIP imagery would be enhanced to provide annually updated one-meter resolution orthoimagery over all states except Hawaii and Alaska.[63] The USGS program would also collect one-foot resolution imagery every three years for 50% of the U.S. land mass (except Alaska, which would get one-foot resolution imagery only over densely populated areas). The program would include an option for states to "buy up," or enhance, any or all of the remaining 50%. The program would also provide 50% matching funds for partnerships to acquire six-inch resolution imagery over urban areas with at least 1,000 people per square mile as identified by the U.S. Census Bureau.

NSGIC states that statewide GIS coordination councils would specify their requirements through business plans, and that all the data would remain in the public domain, which would address many of the data sharing issues discussed above. In addition, the program calls for appropriate national standards for all data, which is a goal of the FGDC, a partner to NSGIC in the development of IFTN. NSGIC estimates that the program would cost $1.38 billion during the first 10 years, and argues that this would save $120 million over the 10-year period by reducing the number of contracts, contracting for larger areas, reducing overhead, and reducing other costs associated with current efforts.[64]

Advancing the National Spatial Data Infrastructure: The NSGIC Perspective

NSGIC considers the 50-States Initiative as one of the crucial components needed to build the NSDI and to bring consistency of geospatial information and parity to each of the states.[65] NSGIC also considers that IFTN is the first of several initiatives creating "core data layers," or baseline data programs, required to meet federal, state, and local needs.[66] NSGIC suggests that the NGAC be an interim step in the governance structure for NSDI, and indicates that the national effort to govern and coordinate the geospatial enterprise should not stifle the states from customizing aspects of the NSDI to suit their own needs:

> the federal government must not dictate the actions of state and local governments, nor should state governments dictate those of local government. However, each level of government can exert a strong influence on subordinate levels by making funding contingent on compliance with the policies and standards it establishes.[67]

NSGIC further argues that funding the resulting collaboration and compliance could be modeled on the federal highway program.[68]

A NATIONAL GIS?

In early 2009, several proposals were released calling for efforts to create a national GIS,[69] or for renewed investment in the national spatial data infrastructure, or even to create a "NSDI 2.0."[70] The release of these proposals coincided with deliberation of major legislation to stimulate the U.S. economy through massive spending on the nation's infrastructure, among other things, that eventually passed as P.L. 111-5, the American Investment and Recovery Act of 2009 (ARRA). The language in the proposals attempted to make the case for considering such investments part of the national investment in critical infrastructure, both by directly supporting these national GIS and geospatial efforts, but also via secondary effects. For example, one proposal indicated that organizations rebuilding roads, bridges, and schools need updated online information networks "to rebuild in a smart, efficient, environmentally conscious and sustainable way."[71] Another proposal touted a national GIS as a tool to speed economic recovery, which should also "leave the country with a public utility, a modern geospatial information system, that itself can become a foundation for new generations of industries and technologies in the future."[72]

The timing of the these proposals and the language linking their purported benefits to national economic recovery were clearly intended to take advantage of the Obama Administration's and Congress's deliberations on economic stimulus funding. None of the proposals was included in ARRA, but their call for efforts to build a "national" GIS, or a new version of the NSDI, or for an investment in a national spatial data infrastructure, raises questions about the current efforts to build the NSDI. Efforts to construct the NSDI began in 1994 with Executive Order 12906, or even earlier when OMB revised Circular A-16 in 1990 to establish the Federal Geographic Data Committee. The more recent proposals imply that efforts which began over 15 years ago and continue today are not sufficiently national in scope, planning, coordination, sharing, or implementation, despite the existence of the FGDC, NSGIC, or other organizations such as MAPPS or COGO that are forums for organizations concerned with national geospatial issues.

The National Geospatial Advisory Committee recommended that OMB and FGDC strengthen their enforcement of Circular A-16 and Executive Order 12906; however, enforcement alone may not be sufficient to meet the current challenges of coordination and data sharing. For example, OMB Circular A-16 was last revised in April 2002, prior to the creation of the Department of Homeland Security (DHS). The current membership of FGDC does not include DHS, which itself has a significant interest in geospatial information. The example of DHS raises the broader issue of data sharing and coordination of geospatial information collected for civilian versus national security purposes in the post-September 11 era.

Congress may wish to consider how a national GIS or geospatial infrastructure would be conceived, perhaps drawing on proposals for these national efforts as described above, and how they would be similar to or differ from current efforts. Congress may also wish to examine its oversight role in the implementation of OMB Circular A-16, particularly in how federal agencies are coordinating their programs that have geospatial components. In 2004, GAO acknowledged that the federal government, through the FGDC and Geospatial One-Stop project, had taken actions to coordinate the government's geospatial investments, but that those efforts had not been fully successful in eliminating redundancies between agencies. As a result, federal agencies were acquiring and maintaining potentially duplicative data sets and systems.[73] Since then, it is not clear whether federal agencies are now successfully coordinating among themselves and measurably eliminating unnecessary duplication of effort. An additional challenge is how Congress oversees the federal geospatial enterprise when so much government information has a geospatial component, and many departments and agencies are actively involved in acquiring and using geospatial data for their own purposes.

APPENDIX. LIST OF ACRONYMS

ARRA	American Recovery and Reinvestment Act of 2009
APZ	Accident Potential Zone
BLM	Bureau of Land Management
BRAC	Base Realignment and Closure
COGO	Coalition of Geospatial Organizations
CZ	Clear Zone
DFIRMs	Digital Flood Insurance Rate Maps
FEMA	Federal Emergency Management Agency
FGDC	Federal Geographic Data Committee
FIRMs	Flood Insurance Rate Maps
GAO	Government Accountability Office
GIS	Geographic Information Systems
GPS	Global Positioning System
IFTN	Imagery for the Nation
LIDAR	Light Detection and Ranging
MAPPS	Management Association for Private Photogrammetric Surveyors
NGAC	National Geospatial Advisory Committee
NGP	National Geospatial Program
NILS	National Integrated Land System
NRC	National Research Council
NSDI	National Spatial Data Infrastructure
NSGIC	National States Geographic Information Council
OMB	Office of Management and Budget
USGS	U.S. Geological Survey
USFS	U.S. Forest Service
VGI	Volunteered Geographic Information

ACKNOWLEDGMENTS

Paul Schirle, Geographic/Geospatial Systems (GIS) Analyst, and Jan Johansson, Data Librarian, both of the CRS Knowledge Services Group, made significant contributions to this report.

End Notes

[1] The development and commercial availability of Global Positioning System (GPS) data and the integration of these data with digital maps has led to the popular handheld or dashboard navigation devices used daily by millions.

[2] The National Geospatial Advisory Committee, *The Changing Geospatial Landscape*, January 2009, p. 10, http://www.fgdc.gov/ngac/NGAC%20Report%20-%20The% 20 Changing % 20 Geospatial%20 Landscape. pdf. Hereafter referred to as NGAC, *The Changing Geospatial Landscape*, January 2009.

[3] NGAC, *The Changing Geospatial Landscape*, January 2009, p. 9.

[4] Prepared statement of Rep. Adam Putnam, Chair, U.S. Congress, House Committee on Government Reform, Subcommittee on Technology, Information Policy, Intergovernmental Relations and the Census, *Geospatial Information: A Progress Report on Improving our Nation's Map-related Data Infrastructure*, 108th Cong., 1st sess., June 10, 2003, H. Hrg. 108-99 (Washington: GPO, 2004).

[5] GAO (2004), p. 11.

[6] U.S. Geological Survey, *Geographic Information Systems*, http://egsc.usgs.gov/isb/ pubs/gis _poster/#what.

[7] National Research Council, *Successful Response Starts With a Map: Improving Geospatial Support for Emergency Management*, Washington, DC, 2007, p. 15.

[8] New York State Department of Environmental Conservation, Center for Technology in Government, *Sharing the Costs, Sharing the Benefits: the NYS GIS Cooperative Project*, Project Report 95-4, Albany, NY, 1995, p. 7, http://www.ctg.albany.edu/ publications/ reports/ sharing _the_costs/sharing_the_costs.pdf.

[9] For example, thousands of amateur geospatial enthusiasts are forming mapping parties, using personal navigation devices to create their own street maps. See http://www.OpenStreetMap.org. Information derived from such groups is referred to as volunteered geographic information (VGI).

[10] A projection is a mathematical means of transferring information from the Earth's three-dimensional, curved surface onto a two-dimensional map or computer screen.

[11] U.S. General Accounting Office, *Geospatial Information: Better Coordination Needed to Identify and Reduce Duplicative Investments*, GAO-04-703, June 23, 2004, p. 13. Hereafter referred to as GAO (2004). GAO became the Government Accountability Office effective July 7, 2004.

[12] Ibid.

[13] National Research Council, *Successful Response Starts With a Map,* 2007, p. 3.

[14] Section 575 of P.L. 103-325 requires the Director of FEMA to assess the need to revise and update all floodplain areas and flood risk zones identified.

[15] For more information on the flood map modernization initiative, see CRS Report R40073, *FEMA Funding for Flood Map Modernization*, by Wayne A. Morrissey.

[16] National Research Council, *Mapping the Zone: Improving Flood Map Accuracy*, Washington, DC, 2009, p. 38. Hereafter referred to as NRC, Mapping the Zone.

[17] See box above for definition.

[18] North Carolina instigated a state-wide LIDAR program, in part, to improve the accuracy of floodplain maps in the wake of hurricane Floyd in 1999. As a result, the state has nearly complete LIDAR coverage.

[19] NRC, *Mapping the Zone*, Recommendations.

[20] See MAPPS, at http://www.mapps.org/.

[21] For a list of the COGO member organizations, see http://www.urisa.org/cogo.

[22] CBS Broadcasting, Inc., "Sayre fire reaches 85 percent containment," November 19, 2008, at http://cbs2.com/local/ brush.fire.Sylmar.2.865252.html.

[23] Mathew Bloch and Janet Roberts, "Mapping Foreclosures in the New York Region," *New York Times*, May 15, 2009, at http://www.nytimes.com/interactive/2009/05/15/nyregion/0515-foreclose.html.
[24] John P. Harding, Eric Rosenblatt, and Vincent W. Yao, "The Contagion Effect of Foreclosed Properties," *Social Science Research Network Working Paper*, July 15, 2008.
[25] Cited in U.S. General Accounting Office, *Geographic Information Systems: Challenges to Effective Data Sharing*, GAO-03-874T, June 10, 2003, p. 5. Hereafter referred to as GAO (2003). The *2006 Annual Report* from the Federal Geographic Data Committee claims that 80%-90% of government information has a spatial component.
[26] The National Geospatial Advisory Committee, *A National Geospatial Strategy: Recommendations for the 2008-2009 Presidential Transition Team*, at http://www.fgdc.gov/ngac/ngac-transition-recommendations-10-16-08.pdf.
[27] U.S. Department of Labor, Employment and Training Administration, at http://www.doleta.gov/brG/Indprof/geospatial_profile.cfm (viewed May 14, 2009).
[28] GAO (2004), p. 11.
[29] See the USGS National Geospatial Program, at http://www.usgs.gov/ngpo/index.html.
[30] Available at http://www.fgdc.gov/library/whitepapers-reports/annual%20 reports/2007/ index_.html.
[31] The website is called geodata.gov and is available at http://gos2.geodata.gov/wps/portal/gos.
[32] See USGS, *The National Map*, at http://nationalmap.gov/index.html.
[33] A brief description of the program, and a link to a list of the liaisons, is provided at http://www.usgs.gov/ngpo/ ngp_liaisons.html.
[34] See http://www.usgs.gov/ngpo/ngp_liaisons.pdf for a list of states and their assigned liaisons.
[35] Telephone conversation with Vicki Lukas, Chief, NGP Partnerships, USGS, Reston, VA, May 21, 2009.
[36] Ibid.
[37] National Research Council, *National Land Parcel Data: A Vision for the Future*, Washington, DC, 2007, p. 2. Hereafter referred to as NRC, National Land Parcel Data.
[38] NRC, National Land Parcel Data.
[39] NRC, National Land Parcel Data, p. 7.
[40] NRC, *Mapping the Zone*, Summary.
[41] For more detail on funding for the program, see CRS Report R40073, *FEMA Funding for Flood Map Modernization*, by Wayne A. Morrissey.
[42] GAO (2003), p. 6.
[43] Circular A-16, at http://www.whitehouse.gov/omb/circulars/a016/a016_rev.html.
[44] The Department of Defense also administers a significant amount of land.
[45] See http://www.blm.gov/wo/st/en/prog/more/nils.html.
[46] See http://www.geocommunicator.gov/GeoComm/index.shtm.
[47] E-mail from John P. Donnelly, National Atlas of the United States, USGS, Reston, VA, February 4, 2009.
[48] H.R. 5532 and S. 3043. Neither version of the bill saw action in the 110th Congress.
[49] "Legislators return with FLAIR," *GEO World*, May 2008, p. 15.
[50] Ibid.
[51] The Digital Coast is a NOAA-led effort envisioned as a an information delivery system for coastal data, as well as the training, tools, and examples needed to turn data into useful information. See http://www.csc.noaa.gov/digitalcoast/ index.html.
[52] See NGDC website, at http://www.fgdc.gov/ngac/index_ html/?searchterm =advisory%20 committee. The committee is sponsored by the Department of the Interior under the Federal Advisory Committee Act.
[53] NGAC, *The Changing Geospatial Landscape*, January 2009, p. 12.
[54] Ibid.
[55] See http://www.fgdc.gov/ngac/ngac-transition-recommendations-10-16-08.pdf.

[56] The NGAC recommendations further specify that an "Urgent Path" forward should include (1) Imagery for the Nation; (2) National Land Imaging Program; and (3) National Land Parcel data.
[57] 13 U.S.C. § 9 and §13. See also U.S. Census Bureau, at http://www.census.gov/privacy/ data_protection/ federal_law.html.
[58] For example, the website for the City of Greeley, CO, property information map, identifies names and addresses, the underlying street map and orthoimagery, together with other information such as school districts and even the nearest fire hydrant. See http://gis.greeleygov.com/origin/propinfo.html.
[59] Telephone conversation with Anne Miglarese, Chair, National Geospatial Advisory Committee, May 26, 2009.
[60] National States Geographic Information Council Strategic Plan 2009-2011, at http:// www.nsgic.org/resources/ strategicplan.pdf.
[61] E-mail from Learon Dalby, NSGIC President 2008-2009, March 11, 2009.
[62] See NISGIC, Imagery for the Nation, at http://www.nsgic.org/ hottopics/imagery for the nation.cfm.
[63] Imagery would be updated once every three years in Hawaii. The USGS program would produce one-meter imagery for Alaska once every five years.
[64] CRS did not review the basis for NSGIC's cost analysis, nor examine the cost benefit analysis completed for the IFTN in July 2007.
[65] NSGIC, Strategic Framework for the NSDI, at http://www.nsgic.org/resources/ strategic_framework _ NSDI _NSGIC.pdf.
[66] Ibid; for example, NSGIC suggests that Imagery for the Nation should probably be followed by Elevation for the Nation, Transportation for the Nation, Cadastral for the Nation, and so on.
[67] NSGIC, The States' Perspective on Advancing the National Spatial Data Infrastructure, October 10, 2008.
[68] NSGIC, Strategic Framework for the NSDI.
[69] These proposals are broader than what is currently envisioned as The National Map, under the USGS.
[70] See, for example, the following: *A Proposal for National Economic Recovery: An Investment in Geospatial Information Infrastructure Building a National GIS*, at http://www.gis.com/gisnation/pdfs/ national_economic_recovery.pdf; *A Proposal for Reinvigorating the National Economy Through Investment in the US National Spatial Data Infrastructure*, at http://www.cast.uark.edu/nsdi/nsdiplan.pdf; and *A Concept for American Recovery and Reinvestment, NSDI 2.0: Powering our National Economy, Renewing our Infrastructure, and Protecting our Environment*, at http://www.nsdi2.net/ NSDI2ProposalFor American RecoveryAndReinvestment_V1_4.pdf.
[71] A Concept for American Recovery and Reinvestment, NSDI 2.0: Powering our National Economy, Renewing our Infrastructure, and Protecting our Environment, p. 2.
[72] *A Proposal for National Economic Recovery: An Investment in Geospatial Information Infrastructure Building a National GIS.*
[73] GAO (2004), p. 19.

Chapter 2

ISSUES REGARDING A NATIONAL LAND PARCEL DATABASE[*]

Peter Folger

SUMMARY

The federal government's efforts to coordinate its geospatial activities, through the Federal Geographic Data Committee (FGDC) and the development of the National Spatial Data Infrastructure (NSDI), include a strong emphasis on land parcel data. Land parcel databases (or cadastres) describe the rights, interests, and value of property. Ownership of land parcels is an important part of the legal, financial, and real estate system of a society. The Bureau of Land Management (BLM) is assigned the role of lead agency coordinating land parcel data for federal lands, and is responsible for performing cadastral surveys on all federal and Indian lands. According to BLM, "Cadastral surveys are the foundation for all land title records in the United States and provide federal and tribal land managers with information necessary for the management of their lands."

Although BLM is steward of federal land parcel data and coordinator for cadastral data under FGDC, a 2007 National Research Council (NRC) report

---

[*] This is an edited, reformatted and augmented version of Congressional Research Service publication, Report R40717, dated July 22, 2009.

found that a coordinated approach to federally managed parcel data did not exist. Legislation that addresses some of the issues for creating a national cadastre has been introduced in the 111th Congress (H.R. 1520, the Federal Land Asset Inventory Reform Act of 2009). Similar bills were introduced in previous Congresses, but were not enacted. In addition, the E-Government Act of 2002 (P.L. 107-347) contains provisions that specifically address reducing data redundancy and promoting collaboration and use of standards for government geographic information. If the E-Government Act was reauthorized, it could also include language establishing a national cadastre. Coordinating all land parcel data, the bulk of which is produced for local and regional needs on non-federal lands, remains even more of a challenge.

Why a national land parcel database? The National Geospatial Advisory Committee (NGAC) observed that the federal government's land parcel data is missing an arrangement for acquiring the detailed property-related data necessary to make decisions during times of emergency, such as a natural disaster. In addition to emergency response to disasters, other perceived needs for a national land parcel database include responding to the home mortgage foreclosure crisis, dealing with wildfires, managing energy resources on federal lands, dealing with the effects of climate change, and possibly more.

Both administrative and legislative options have been proposed to achieve the vision for a land parcel database described in the 2007 NRC report: a distributed system of land parcel data housed with the appropriate data stewards but accessible through a web-based interface. Some recommend that the Office of Management and Budget (OMB) and the Department of the Interior take a stronger hand in enforcing the requirements of OMB Circular A-16 and Executive Order 12906, which created the FGDC and instigated efforts to create the NSDI. NGAC, for example, also recommended establishing a Geographic Information Officer within each federal department or agency, and establishing a geospatial leadership and coordination function in the Executive Office of the President. The NRC recommended the creation of both a federal land parcel coordinator and a national land parcel coordinator. The first would be responsible for federal lands and property; the second would coordinate parcel data from all sources, both public and private lands. A truly national land parcel cadastre would likely require strong partnerships between the federal government and state and local governments.

INTRODUCTION

This report provides a summary of some of the issues regarding the creation of a national land parcel database, or cadastre.[1] The report identifies some of the perceived needs for a national cadastre, legislative and administrative options that could lead to a national land parcel database, and some of the challenges and concerns. The report also summarizes and briefly discusses recommendations in a 2007 National Research Council (NRC) report that concluded "... a national approach is necessary to provide a rational and accountable system of property records."[2] The NRC report described why a national approach is needed, identified challenges to creating a national cadastre, and offered specific recommendations for achieving its vision: a distributed system of land parcel data housed with the appropriate data stewards but accessible through a web-based interface.[3]

Legislation that addresses some of the issues for creating a national cadastre has been introduced in the 111th Congress: H.R. 1520, the Federal Land Asset Inventory Reform Act of 2009. Similar bills were introduced in the 110th and 109th Congresses, but were not enacted.

For more information on geospatial information generally, see CRS Report R40625, *Geospatial Information and Geographic Information Systems (GIS): Current Issues and Future Challenges*.

WHY A NATIONAL LAND PARCEL DATABASE?

Geospatial information, including land parcel data,[4] is increasingly produced by private sector and other non-federal government sources. Consequently, the federal government's role has shifted from producing geospatial data to coordinating efforts, facilitating partnerships, and managing the vast amounts of geospatial information.[5] According to the National Geospatial Advisory Committee (NGAC),[6] the shift in geospatial data production from the federal government to the private sector and state and local governments has created an "... urgent need to reexamine the relationships between data providers and users to establish a fair and equitable geospatial data marketplace that serves the full range of applications."[7] As an example, NGAC noted that the Census Bureau had to develop a duplicate version of street centerlines in preparation for the 2010 Census because it could not take advantage of the existing commercial data.[8] Further, "critical information about the use, value and ownership of property is needed by FEMA, the Forest Service, and HUD, for emergency preparedness or response in times of hurricanes or wildfires—or even to monitor the current foreclosure problems."[9]

Current Status

The federal government's efforts to coordinate its geospatial activities, through the Federal Geographic Data Committee (FGDC) and the development of the National Spatial Data Infrastructure (NSDI), include a strong emphasis on land parcel data. For example, the cadastral data theme is one of the seven fundamental data themes of the NSDI framework. Within the FGDC, the Bureau of Land Management (BLM, in the Department of the Interior) is assigned the role of lead agency coordinating land parcel data for federal lands. According to BLM, it is responsible for performing cadastral surveys on all federal and Indian lands: "Cadastral surveys are the foundation for all land title records in the United States and provide federal and tribal land managers with information necessary for the management of their lands."[10]

Despite the BLM role as steward of federal land parcel data and coordinator for cadastral data under FGDC, NRC found that a coordinated approach to federally managed parcel data did not exist. The National Integrated Land System (NILS) [11]—a joint project between BLM and the U.S. Forest Service (USFS, in the Department of Agriculture)—is the closest thing to a coordinated program "... but it remains much more of a set of technologies than a source of parcel data."[12] Coordinating all land parcel data, the bulk of which is produced for local and regional needs, remains even more of a challenge.

Perceived Need

The National Geospatial Advisory Committee (NGAC) was formed in early 2008 to provide advice and recommendations to the FGDC on management of federal geospatial programs. NGAC observed that the federal government's need for land parcel data is missing an arrangement for acquiring the detailed property-related data necessary to make decisions during times of emergency. In addition to emergency response related to natural disasters, other perceived needs for a national land parcel database at the federal level include responding to the home mortgage foreclosure crisis, dealing with wildland fires, and managing extractive energy resources on federal lands.[13] Other aspects of natural resources management on federal lands could be included as well, such as monitoring the effects of climate change, and the efficacy of measures taken to mitigate or adapt to such effects.

Natural Disasters

Disasters are often cited as a compelling reason to establish a national land parcel database: "Land-parcel data, one of the framework themes, are essential in managing disasters and in assessing damage, along with building footprints and the locations of infrastructure (power, telecommunications, water, sewage, and steam-heating networks)."[14] The attacks of September 11, 2001, and hurricanes Katrina and Rita in 2005, underscored for many the need for rapid access to land ownership data to help guide emergency response, especially when a disaster crosses multiple jurisdictions or extends beyond the boundaries of a community and the immediate knowledge of local responders. The land parcel data useful to emergency responders may exist, but may also be difficult to access:

> Data on the ownership of land parcels, or cadastral data, provide a particular and in some ways extreme example of the problems that currently pervade the use of geospatial data in emergency management. Vast amounts of such data exist, but they are distributed among tens of thousands of local governments, many of which have not invested in digital systems and instead maintain their land-parcel data in paper form. As with many other data types, it is not so much the existence of data that is the problem, as it is the issues associated with rapid access.[15]

Several NRC reports noted that a national partnership for assembling land parcel data would provide major benefits for managing federal assistance to local programs, many of which are associated with the U.S. Department of Housing and Urban Development (HUD).[16] According to the NRC, parcel-level data would help HUD meet its strategic goals, such as increasing home ownership opportunities, promoting affordable housing, and ensuring equal opportunities in housing. NRC further contended that "the existence of national land parcel data would provide HUD with data it needs for effective management of grants and would have avoided the critical time wasted gathering parcel data piecemeal in the wake of these recent hurricanes."[17]

Home Mortgage Foreclosure Crisis

In addition to natural disasters, land parcel data are being used for responding to the housing market collapse that began in 2008. The FGDC Cadastral Subcommittee noted that parcel data provide added value to the mortgage and property information collected by the federal government under the Home Mortgage Disclosure Act (HMDA).[18] HMDA was enacted in 1975 to assist government regulators and the private sector with the monitoring of anti-discriminatory practices.[19] According to the FGDC Cadastral Subcommittee

> While HMDA data provide a snapshot in time of a mortgage transaction, local government parcel data provide current information at the individual parcel level that allows other information such as utility shutoffs, code violations and undelivered mail to be tied to a common unit, the parcel. Parcel data make it possible to relate disparate data together to get a complete picture of individual mortgage and housing conditions. Parcel data also provide the connection to local governments, which can provide community context and engage those most affected by mortgage crisis events.[20]

The Cadastral Subcommittee likened the distressed housing market to a contagious disease, tending to affect some communities while leaving others intact. By adding parcel data to existing information available under the authority of HMDA, data analyses could identify "hot spots" in a pending foreclosure crisis, and possibly even provide sufficient information for a national early warning system for financially distressed housing and mortgage markets.[21]

In one case, GIS and land parcel data were used to identify and analyze the extent of home foreclosures, and to use the results of that analysis to apply for Community Development Block Grants (CDBG) to convert foreclosed properties into low-income housing.[22] It could be asserted that these types of land parcel data, made available to federal agencies such as HUD, could also be used to track the effects of programs like CDBG to ameliorate the foreclosure crisis. This type of use of land parcel data arguably underscores a need for a national land parcel database to track the effectiveness of federal agency programs in national efforts, such as coping with the home foreclosure crisis.

Wildfires

The FGDC Cadastral Subcommittee formed a Wildland Fire Project Team, at the request of the National Interagency Fire Center, together with representatives from BLM, USFS, and the U.S. Geological Survey, state representatives, and others to prepare for the 2007 fire season.[23] The goal was to identify contacts for parcel data in priority counties throughout the West, and acquire and have as much parcel data as possible pre-deployed to support analyses of and responses to wildfires. The project was also intended to foster coordination between the cadastral community and the wildland fire community to identify the cadastral data needs to support planning for, response to, and mitigation of wildfires.

According to a 2007 report by the Cadastral Subcommittee, "... structures located within the wildland-urban interface comprise a very substantial portion of values commonly threatened by wildland fires. GIS parcel data from local and state government provide effective and accurate means to identify and map general structure locations with associated values."[24] These data are used to provide rapid analyses and wildfire suppression strategies by quantifying the significant resource values most threatened by a fire.

Following the very active 2007 fire season, the Cadastral Subcommittee observed that to increase the efficiency and sustainability of the effort, several changes were needed:

- increasing state-level participation and involvement to help build a single state contact for parcel information;
- merging the point-of-contact information with the 50-States Initiative[25] into a single data and point of contact resource;
- expanding the use of pre-deployed parcel data to support other aspects of emergency response and reduce duplicative parcel inventory efforts; and
- obtaining federal assistance to work with states that work with counties to complete and standardize parcel data systems.[26]

The wildland fire project may represent an example of how making land parcel data available, from the local and state level through the federal level, could serve multiple stakeholders who benefit from access to the data. Whether this example can be expanded to all states susceptible to wildfires, or to the entire country in a multihazard approach, remains an open question.

Energy Resources

The FGDC Cadastral Subcommittee identified a need for accurate survey boundaries and land ownership information (i.e., land parcel information) for management of the life cycle of energy development from prospect to production to remediation.[27] For western states, where much of the nation's onshore energy production occurs, the Public Land Survey System (PLSS) is the primary cadastral framework, supported by BLM's Cadastral Survey Program and represented in a digital format by the Geographic Coordinate Data Base (GCDB).[28] The Cadastral Subcommittee proposed a set of elements comprising an "energy core" set of information that could be provided by land parcel data producers in energy production areas—referenced to the cadastral framework of the GCDB—and could lend efficiency and accuracy at each stage of energy production activities: application, permit, monitoring, and reclamation activities. As with other applications, such as wildfire support, the Cadastral Subcommittee underscored the need to embrace and apply consistent cadastral framework standards to parcel data.[29]

In western states, energy resources are commonly exploited on a variety of lands: federally managed surface and subsurface lands; state, county, tribal, or privately owned lands; and split estates where the surface lands may be privately owned but the minerals are federally managed (or vice-versa). The Cadastral Subcommittee observed that "in all of these cases it is essential to build a seamless presentation of surface and subsurface ownership to correctly manage and exploit energy resources."[30] It might be argued that similar needs arise for other parts of the country, such as parts of Pennsylvania, New York, and West Virginia, where exploration and development of potentially huge natural gas deposits in black shales is occurring. Also, if Congress enacts climate change legislation, such as a cap-and-trade system, deployment of capture, transportation, and underground storage of carbon dioxide from industrial facilities could rapidly expand across the nation. Efficient management of surface and subsurface lands and resources for carbon dioxide capture and storage may also benefit from the type of seamless presentation of land parcel data recommended by the Cadastral Subcommittee.

Climate Change

In addition to its potential application to carbon dioxide capture, transportation, and storage mentioned above, a national land parcel system could have other benefits related to mitigating climate change. Legislation intended to deal with anthropogenic climate change, such as under a cap-and-trade program to reduce greenhouse gas emissions, passed the House on June

26, 2009 (H.R. 2454), and the Senate is expected to take up legislation of a similar scope. If enacted, the legislation would have far-reaching effects on the U.S. energy and economic infrastructure, with the goal of reducing the impact of climate change on the nation's farmlands, forests, rivers and streams, coastlines, and ecosystems, as well as human health and well-being. It could be argued that measuring the effectiveness of the emissions-reduction program would depend, in part, on a precise understanding of the ecosystem, agricultural, forest, coastline, and other boundaries that are anticipated to change in response to climate change. Land parcel data potentially could be useful for such types of analyses.

ADMINISTRATIVE AND LEGISLATIVE OPTIONS

Executive Order 12906 and OMB Circular A-16 created the FGDC and instigated efforts to create the NSDI, which includes cadastral data as one of the seven fundamental themes. The FGDC designated BLM as the steward for the federal land parcel data and the coordinator of cadastral data generally, and BLM sponsors the FGDC Subcommittee for Cadastral Data. The Cadastral Subcommittee has made significant progress in the establishment of standards and coordination of cadastral data, according to the NRC.[31] Some contend that data standards and specifications are no longer an issue or a barrier to implementation of a national land parcel database.[32] In addition to administrative imperatives contained within EO 12906 and Circular A-16, legislation such as the E-Government Act of 2002 (P.L. 107-347) contains provisions that specifically address reducing data redundancy and promoting collaboration and use of standards for government geographic information.[33] Despite nearly 20 years of effort at coordinating geospatial information and land parcel data, however, the NRC observed:

> ... one could conclude that the United States has a comprehensive approach to parcel data. However, a detailed analysis of the situation suggests the opposite.... It is difficult to ascertain the status of parcel data within the various federal agencies, and it appears that none of the federal land management agencies have a comprehensive and complete parcel data set for the lands they manage.... There is also evidence that many federal agencies that do not manage lands are acknowledging that they need parcel data to fulfill their missions and, in the absence of a national means to access the data nationwide, are creating data sets to meet their particular needs, often

without coordination with other federal agencies that may have needs for the same or similar data.[34]

Administrative Options

OMB revised Circular A-16 in 2002 and added the Deputy Director of Management, OMB, as vice-chair of the FGDC to serve with the Secretary of the Interior. The revised leadership structure is seen, in part, as an attempt to improve the coordination and oversight of the participating agencies by giving OMB a defined role. Some argue, however, that OMB could take a stronger role in FGDC through more active enforcement. Thus, an administrative option for enforcing a national land parcel database, at least for the federal lands, is to enforce Circular A-16 more rigorously. This would likely mean that OMB would take a true oversight and coordination role and enforce compliance with A-16 through its power to affect the budgets of the participating departments and agencies. The National Geospatial Advisory Committee (NGAC) recommended this action, and further recommended that the Administration establish a Geographic Information Officer within each department or agency with responsibility under FGDC.[35] NGAC also recommended the establishment of a geospatial leadership and coordination function in the Executive Office of the President, which would elevate the profile of the geospatial enterprise within the Administration and presumably signal a higher priority for coordinating geospatial activities in the federal government.

Legislative Options

H.R. 1520, the Federal Land Asset Inventory Reform Act of 2009

On March 16, 2009, Representative Kind introduced H.R. 1520, the Federal Land Asset Inventory Reform Act of 2009, which would require the Secretary of the Interior to develop a multipurpose cadastre of federal "real property." The legislation defines cadastre as an inventory, and defines federal "real property" as land, buildings, crops, forests, or other resources still attached to or within the land, improvements or fixtures permanently attached to the land, or structures on it. The bill requires the Secretary to coordinate with the FGDC pursuant to OMB Circular A-16, integrate the activities under the legislation with similar cadastral activities of state and local governments,

Issues Regarding a National Land Parcel Database   45

and participate in establishing standards and protocols that are necessary to ensure interoperability of the geospatial information of the cadastre for all users. Similar legislation was introduced in the Senate and House in the 110th Congress.[36]

The legislation includes a provision for a cost-sharing arrangement with states to include any nonfederal lands within a state in the cadastre (§2(b)). The cost-sharing agreement would presumably provide an incentive for the states to share their land-parcel data—namely the federal government would pay up to half the cost—although it is unclear whether the cost incentive alone is enough to compel states to pay the remaining share for a cadastre focused on federal real property. The total cost to the federal government would likely depend, in part, on the degree of participation by the states and the extent and status of their land parcel data. The overall cost of the bill is not clear, but the legislation would require the Secretary to report on the total amount expended on federal land parcel activity in FY2008, and to estimate the cost savings that would be achieved by eliminating or consolidating duplicative real property inventories by creating the multipurpose cadastre.

Sensitive Information

The National Geospatial Advisory Committee recommends revising "... restrictive statutory language as it pertains to non-sensitive address data in Title 13 U.S. Code and to 'geospatial' data in Section 1619 of the 2008 Farm Bill."[37] In Title 13, Congress delegates responsibility for conducting the decennial Census to the Secretary of Commerce. The law contains provisions for not disclosing or publishing private information that identifies an individual or business (Sections 9 and 214 of Title 13). The Census Bureau is forbidden to publish any private information—such as names, addresses, or telephone numbers—that identifies an individual or business.[38] If a legislative proposal to amend portions of Title 13 was introduced to make geospatial data collected by the Census Bureau more accessible (e.g., for use in a national land parcel database), it could raise issues about the privacy of personal data collected by the federal government. The NRC recommended that Congress and the Bureau of the Census explore various policy options that would allow digital data on building addresses and geographical coordinates to be placed in the public domain while maintaining important privacy protections. (See NRC recommendation 6, below.)

Section 1619 of the 2008 Farm Bill[39] prohibits disclosure of geospatial information about agricultural land or operations when the information is provided by an agricultural producer or owner of agricultural land, and maintained by the Secretary of Agriculture. Certain exceptions, contained in that section, apply to the prohibition. NGAC has taken the position that the statutory language could be revised to enhance the value of the geospatial data, which could then be included in a national land parcel database, while not compromising privacy.[40] For example, the boundaries of fields could be separable elements of a database, not tied to proprietary information about program participation and payments. Boundary information, by itself, might be used for land use planning, conservation, resource management, or possibly other types of applications.

Reauthorizing the E-Government Act

Section 216 of P.L. 107-347, the E-Government Act of 2002, calls for facilitating the development of common protocols for geographic information to promote collaboration and use of standards and to reduce redundancy among federal agencies. Authorization for appropriations under the act expired in FY2007. If the E-Government Act was reauthorized, Section 216 could be expanded to include language for a national cadastre, as proposed in H.R. 1520, for designating Executive Office of the President level leadership for all federal geospatial activities, as recommended by NGAC, or for amending Title 13 of the U.S. Code to enable broader sharing of address data and its inclusion in a national land parcel database.

NRC RECOMMENDATIONS FOR INTEGRATED NATIONAL LAND PARCEL DATA

The NRC made nine recommendations that it believes could lead to a coordinated and integrated national approach to land parcel data, summarized and discussed briefly as follows:[41]

1. Creation of both a federal land parcel coordinator and a national land parcel coordinator. The first would be responsible for federal lands and property; the second would coordinate parcel data from all sources, both public and private. NRC recognizes that BLM is one organizational choice to coordinate the federal land parcel data, and it could serve both roles, but other agencies are also candidates. The Department of Homeland Security (DHS), for example, could establish a national land parcel database as a homeland security issue.

The General Services Administration (GSA) already provides services for all federal agencies. Likewise, the Census Bureau and HUD deal with property issues and need land parcel data to fulfill their missions. NRC recommended that a panel be established to recommend agency leadership. To date, no such panel has been established.[42]

2. FGDC identification of the role of parcel data for the collection and maintenance of other data themes in the overall geospatial infrastructure: buildings and facilities, cultural resources, governmental units, and housing. NRC recommended a systematic review of how these themes would be managed if an integrated national parcel database existed.

3. Development by the federal land parcel coordinator of a single database for land parcels managed by the federal government. This recommendation appears to call for the federal government to house and maintain a single database of federal property, as different from the national land parcel coordinator who would coordinate land parcel data from all sources, which may be housed and maintained in a variety of state, county, local, private, and other databases.

4. Development and oversight by the national land parcel coordinator of a land parcel data business plan for the nation. NRC pointed to the lack of a coordinated federal program for parcel data.

5. Establishment by the Office of the Special Trustee for Tribal lands of an Indian Lands Parcel Coordinator to develop a land parcel database for Indian trust parcels. NRC indicated that this could reduce redundancies and duplication of effort in mapping Indian lands, among other issues related to trust lands.

6. Exploration by Congress and the Bureau of the Census of policy options, including amending Title 13 of the U.S. Code, to allow its digital data on building addresses and their geographic coordinates to be placed in the public domain while maintaining privacy protections.

7. Adoption by the national land parcel coordinator of the 50-States Initiative[43] and require that each state formally establish a state parcel data coordinator. The 50-States Initiative was proposed by the National States Geographic Information Council to develop Statewide Spatial Data Infrastructures (SSDI) for each state. The 50-States Initiative would potentially enable coordination between geospatial data producers and consumers at all levels within the state, and allow the state to share geospatial data with the national geospatial structure envisioned as the NSDI.

8. Development by the national land parcel coordinator of a plan for an intergovernmental funding program for the development and maintenance of parcel data. NRC recognized that the plan must provide financial incentives to local governments that produce and maintain the majority of the parcel data. Additionally, NRC stated that the program would require new funding in addition to existing funding for current federal programs that require parcel data.
9. Requirements that local and state governments make certain aspects of their parcel data available in the public domain, as a prerequisite for participating in federal geospatial programs.

CHALLENGES AND CONCERNS

Several challenges to a coordinated and integrated national approach to land parcel data have been identified, such as confidentiality, cost, collaboration and data sharing, and incentives for state and local governments to participate in a national cadastre. Of the range of potential challenges and concerns, the NRC concluded,

> ... the financial and technical issues are minor compared to the organizational and political ones. With thousands of counties or other governmental entities as potential producers of parcel data, the organizational issues are complex. It is not a simple task to assemble parcel data that span several counties or states. Overcoming organizational boundaries even among federal agencies has been difficult, as evidenced by the fact that there is no single inventory of federal lands.[44]

Several of the legislative and administrative options discussed above address organizational challenges, as do several of the nine NRC recommendations. The NRC also identified political challenges confronting a coordinated and integrated national approach to parcel data: " ... the lack of political will may be the most difficult hurdle of all."[45] NRC lists a range of political challenges:

- Return on investment. Determining how to calculate the benefits and costs of creating a national approach to parcel data is difficult. NRC stated that the real benefits of a nationally integrated system accrue to groups larger than local government agencies seeking improved tax

Issues Regarding a National Land Parcel Database 49

compliance or improved local government efficiency. NRC contended that a national system would result in reduced fraud, fairer tax assessments, more effective emergency management and response, improved economic development, and other benefits.
- Motivation at the local level. What does and could motivate local governments, which manage land parcel systems for local needs, to participate in a national program? According to the NRC, some local governments assume that a national system could never be as accurate as their own data, and that they also fear releasing information to the public domain that the local government paid for.
- Unfunded mandates. The NRC noted that local governments face many budget restrictions, and some distrust the forced sharing of data with nothing tangible in return.
- Private sector benefits. The NRC reported a widespread perception that many private firms are harvesting data collected by local governments for commercial gain, without any perceived benefits flowing back to the local government.
- Other local political realities. The NRC acknowledged that local political leaders may struggle with approving budget requests for large technical projects, such as county participation in a national effort to create an integrated land parcel database, especially when the benefits to the local government are not clear.

Lastly, the NRC concluded that "With more than 3,000 counties, tribes, and other local government entities as potential producers of parcel data, the organizational issues are complex."[46]

Some of these concerns have been echoed by the National States Geographic Information Council (NSGIC); however, NSGIC also embraces the need for better coordination and for a national spatial data infrastructure, which would include a national land parcel component. The states are sensitive to imposing a federal program, however, and are more likely to work in partnership with the federal government. NSGIC recommends its 50-States Initiative to meet the needs of the states while also sharing land parcel data with the national program. The NRC also recommended that a national land parcel coordinator adopt the 50-States Initiative.

The Western Governors' Association (WGA) has also supported federal, state, tribal, and local coordination of GIS activities and encouraged regional, state, and interstate data sharing.[47] Further, WGA recognized that BLM is working with state and local governments to develop current and standardized digital representations of the Public Land Survey System and parcel data, and has referred to this collaboration as the Cadastral National Spatial Data Infrastructure (Cadastral NSDI). The Western Governors called on Congress to provide the funding necessary for BLM to complete, enhance, and maintain the Cadastral NSDI in coordination and partnership with state, tribal, and local governments.[48] One estimate of funding to implement the WGA recommendation is $350 million over three years, followed by a smaller amount in each succeeding year to maintain and enhance a Cadastral NSDI.[49]

End Notes

[1] Cadastre is the map of ownership and boundaries of land parcels.
[2] National Research Council, *National Land Parcel Data: A Vision for the Future*, Washington, DC, 2007, p. 113. Hereafter referred to as NRC, *National Land Parcel Data*.
[3] Ibid.
[4] Land parcel databases describe the rights, interests, and value of property. The legal boundaries of land parcels are defined in the deed to a property, and are confirmed by survey measurements. Ownership of land parcels is an important part of the legal, financial, and real estate system of a society. See NRC, *National Land Parcel Data*, Introduction.
[5] The National Geospatial Advisory Committee, *The Changing Geospatial Landscape*, January 2009, p. 12. Hereafter referred to as NGAC, 2009.
[6] Its members include federal, state, and local government representatives, private sector representatives, and academics.
[7] NGAC, 2009, p. 12.
[8] This duplication in effort was a result, in part, of prohibitions on disclosing or publishing private information that identifies an individual or business, per Title 13 of the U.S. Code.
[9] NGAC, 2009, p. 12.
[10] U.S. Department of the Interior, Bureau of Land Management, Cadastral Survey Program, at http://www.blm.gov/ wo/st/en/prog/more/cadastralsurvey/program_description.html.
[11] See http://www.blm.gov/wo/st/en/prog/more/nils.html.
[12] NRC, *National Land Parcel Data*, p. 3.
[13] Telephone conversation with Nancy von Meyer, vice president, Fairview Industries, Pendleton, SC, July 20, 2009.
[14] National Research Council, *Successful Response Starts With a Map*, Washington, DC, 2005, p. 38.
[15] NRC, *Successful Response Starts With a Map*, p. 90.
[16] These include National Research Council, *GIS for Housing and Urban Development*, Washington, DC, 2003; and National Research Council, *Procedures and Standards for a Multipurpose Cadastre*, Washington, DC, 1983.
[17] NRC, *National Land Parcel Data*, p. 47.
[18] P.L. 94-200, 12 U.S.C. §§ 2801-2809.
[19] For more information, see CRS Report RL34720, *Reporting Issues Under the Home Mortgage Disclosure Act*, by Darryl E. Getter.
[20] Federal Geographic Data Committee Cadastral Subcommittee Mortgage Study Team, "Land Parcel Data for the Mortgage Crisis: Results of the Stakeholders Meeting," June 30, 2009, p. 4, at http://www.nationalcad.org/data/ documents/Land_parcel_data_for_the_ mortgage_ crisis_-_stakeholders _ meeting _ findings.pdf.

[21] Ibid., p. 6.
[22] Government Technology, *GIS Maps Track Foreclosures in California and Kansas*, April 29, 2009, at http://www.govtech.com/gt/649520.
[23] FGDC Cadastral Subcommittee, "Briefing Paper: Pre-Deploying Parcel Data for Managing Wildland Fires," at http://www.nationalcad.org/data/documents/RavarBrief.pdf.
[24] FGDC Cadastral Subcommittee, "Parcels and Wildland Fire, 2007 Report," January 2008, Preface, at http://www.nationalcad.org/data/documents/Parcels%20and%20Wildland%20Fire%202007%20final%20report.pdf.
[25] For more information about the 50-States Initiative, see NSGIC, at http://www.nsgic.org/hottopics/ fifty_states.cfm.
[26] FGDC Cadastral Subcommittee, "Parcels and Wildland Fire," January 2008, p. 3.
[27] FGDC Cadastral Subcommittee—Energy Workgroup, "The Energy Community and Cadastral Data," May 2006, at http://www.nationalcad.org/data/documents/ The_Cadastral_NSDI_and_the_ Energy_Community.pdf.
[28] For more information on the BLM program, see http://www.blm.gov/wo/ st/en/prog/ more/gcdb.html.
[29] FGDC Cadastral Subcommittee—Energy Workgroup, "The Energy Community and Cadastral Data," May 2006, p. 14.
[30] Ibid., p. 9.
[31] NRC, *National Land Parcel Data*, p. 69.
[32] Telephone conversation with Nancy von Meyer, vice president, Fairview Industries, Pendleton, SC, July 20, 2009.
[33] 44 C.F.R. § 3501 note.
[34] NRC, *National Land Parcel Data*, p. 69.
[35] The National Geospatial Advisory Committee, *A National Geospatial Strategy: Recommendations for the 2008-2009 Presidential Transition Team*, at http://www.fgdc.gov/ngac/ngac-transition-recommendations-10-16-08.pdf.
[36] H.R. 5532 and S. 3043. Neither bill saw action in the 110th Congress.
[37] NGAC, 2009.
[38] 13 U.S.C. § 9 and §13. See also U.S. Census Bureau, at http://www.census.gov/ privacy/data_ protection/ federal_law.html.
[39] P.L. 110-246.
[40] Telephone conversation with Anne Miglarese, Chair, National Geospatial Advisory Committee, May 26, 2009.
[41] NRC, *National Land Parcel Data*, chapter 7.
[42] It should be noted NGAC recommended that immediate action be taken on this recommendation. See National Geospatial Advisory Committee, Summary of Key Decisions/Recommendations from NGAC Meetings, April 2009, at http://www.fgdc.gov/ngac/ngac-summary-key-recommendations-apr-09.pdf.
[43] For more information about the 50-States Initiative, see NSGIC, at http://www.nsgic.org/hottopics/fifty_states.cfm.
[44] NRC, *National Land Parcel Data*, p. 3.
[45] Ibid., p. 108.
[46] NRC, *National Land Parcel Data*, p. 112.
[47] Western Governors' Association, Policy Resolution 09-8, "Collaborative Geographic Data is Part of the Nation's Critical Infrastructure," at http://www.westgov.org/ wga/policy/09/GIS.pdf.
[48] Western Governor's Association, Policy Resolution 09-8, "Collaborative Geographic Data is Part of the Nation's Critical Infrastructure," at http://www.westgov. org/wga/policy/09/GIS.pdf.
[49] Telephone conversation with Nancy von Meyer, vice president, Fairview Industries, Pendleton, SC, July 20, 2009.

In: Geospatial Information and GIS: Background... ISBN: 978-1-61761-432-3
Editor: Sean C. Dallon © 2011 Nova Science Publishers, Inc.

Chapter 3

THE CHANGING GEOSPATIAL LANDSCAPE[*]

National Geospatial Advisory Committee

Practically overnight, access to terabytes of geographical information, much of it in three dimensions, has changed the way people work, live and play. We rely on a host of location-based technologies via our desktop computers, PDAs and even our cell phones. These services fuel a market estimated at $30 billion per year and represent a major information technology growth sector. The primary reasons mainstream commercial applications have emerged are that a wide variety of businesses have taken advantage of investments and policy decisions made by the United States government during the past thirty years, and burgeoning technology innovations. These innovations include the Internet, communications infrastructure, detailed digital mapping, robust data management systems, advancements in modeling the earth's sphere, the creation of a constellation of global positioning system (GPS) satellites, and more.

Enlightened public policies now support shared geospatial technology, thereby fostering a strong international commercial market. For continued benefit to society, it is incumbent upon the nation's policy leaders to understand these points: the government's role in creating and developing these services, how much the landscape has changed during the past 30 years,

---

[*] This is an edited, reformatted and augmented version of A Report of the National Geospatial Advisory Committee, dated January 2009.

and what leaders must do to ensure continued advancement in geospatial technology in the future.

A BRIEF HISTORY OF INFLUENTIAL EVENTS, DIGITAL ROADS, GPS AND LOCATION AWARENESS

The detailed street maps that support Web-based mapping applications and in-car navigation systems can be traced to the innovations made by the Census Bureau approximately forty years ago. Since the initial creation of digital street maps, designed to support the 1970 Decennial Census, the street map data industry has evolved into two multibillion- dollar European companies.

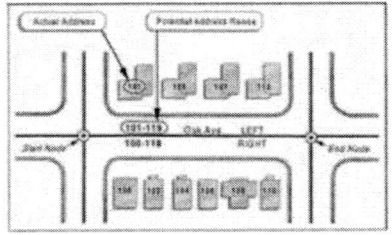

TIGER data (above); early MapQuest

The initial experiments were expanded in the mid 1980s when the Census Bureau teamed up with the US Geological Survey to generate the first nationwide digital street map with address ranges. This became the TIGER system that supported the 1990 Census and forever changed the way we interact with maps. In 1996, MapQuest leveraged these intelligent street maps to build a Web-based system that could determine the geographic location of a street address and display it on a map. MapQuest was an overnight sensation that received 1 million hits in its first 30 days (now 40 million per month). The sale of MapQuest to Aol for $1.1 billion in 1999 represents a landmark in the evolution of the geospatial technology and marks the date when location-based services officially became part of mainstream Internet business.

The need to keep street map and address data current resulted in the creation of Geographic Data Technology (GDT) and Navteq, which have recently been acquired by European companies. GDT was initially purchased by the Belgium company TeleAtlas in 2004, and is now being acquired by

The Changing Geospatial Landscape   55

TomTom, a Dutch personal navigation supplier. Navteq has been purchased by the Finnish telecom giant Nokia for eight billion dollars. The fact that a major telecom company would place that kind of price tag on geospatial data and technology demonstrates the value of these assets and points toward further vertical integration of location-based services, especially on cell phones and PDAs.

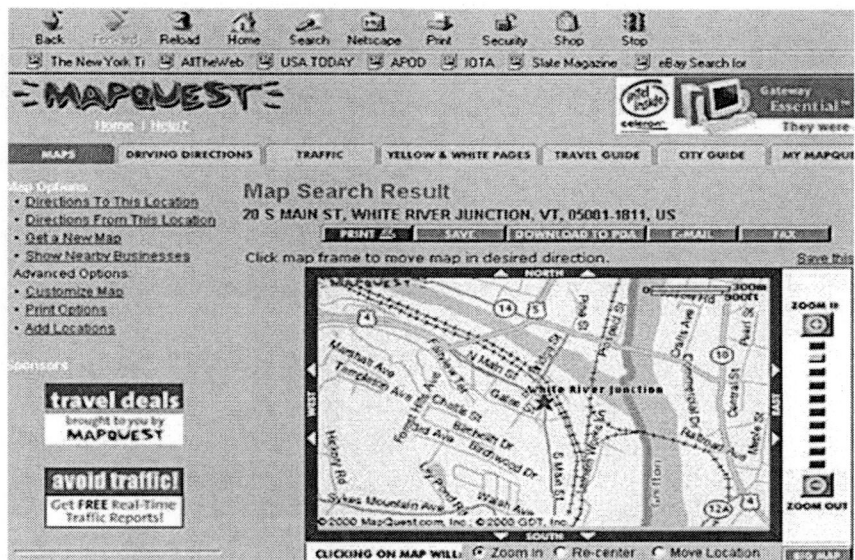

Even though detailed digital street maps provide the basis for spatial search and navigation, they do not actually show consumers their immediate locations. This task is handled by another American innovation: the global positioning system, or GPS. GPS was designed in the mid 1970s to support U.S. Department of Defense missions. In the mid 1990s, the 24 satellites that formed the GPS operational Constellation made it possible to locate geographic coordinates without reference to any landmarks or features on Earth. By recording signals from at least four of the satellites, these GPS receivers were able to determine the X, y and Z coordinates of the receiver anywhere on the Earth's surface or on an aircraft. Since 2000 almost any GPS receiver is able fix a location within a few meters of its actual location.

The accuracy of GPS can be enhanced by a network of land-based survey stations that provide precise coordinates required for surveying. This precision is made possible by the development of a highly accurate model of the earth's shape. A series of enlightened federal policy decisions opened this military system to commercial applications and has spurred a huge new international commercial market. Consequently, creative entrepreneurs have coupled these incredible and inexpensive tools to build hundreds of applications that support the public's insatiable appetite for location-based information.

As the cost of GPS receivers has plummeted, the range of applications has skyrocketed. Personal navigation systems manufactured by GPS technology companies such as Garmin and TomTom represent the integration of digital maps and GPS technology. The demand for navigational assistance has been at the forefront of this trend and has been a major boon to car rental agencies. Furthermore, inexpensive personal navigation systems that cost a few hundred dollars have become popular consumer items.

Some models provide users with task status as well as real-time location information such as traffic conditions, and can even track other people and assets. This tracking capability is now widely deployed to follow the movement of children, employees, criminals, vehicles and even fish. A pet products company sells a GPS dog collar; for a monthly fee, owners can track their pets' locations. The fact that other people can follow your movements (geo-tracking) with or without your permission or knowledge elicits a variety of reactions ranging from comfort to reluctant acceptance to outrage. In fact, some academics have labeled geo-tracking "geo-slavery."

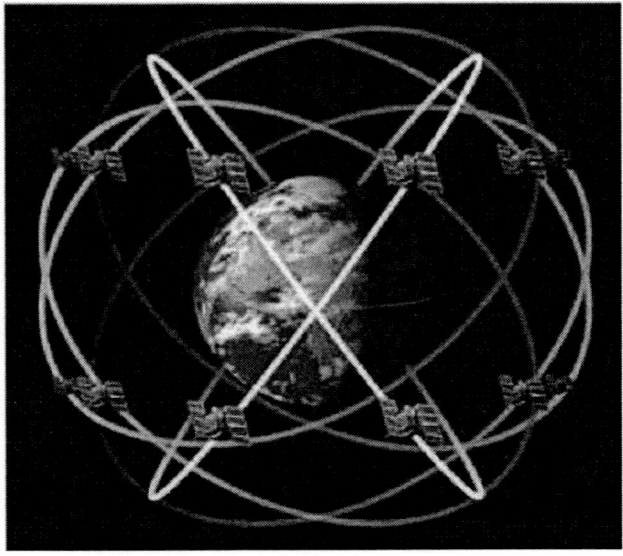

The Global Positioning System satellite network

Personal navigation device

Telecom companies such as Nokia join in a vision of the future that places a high value on accurate geographic information. They plan to embed geospatial technology in the next generation's social psyche in the same way email has become ubiquitous to this generation. An example of such innovation is the Apple iPhone, whose embedded GPS receiver wirelessly accesses the Internet anywhere in the world and integrates its location coordinates with both self-contained and Web-accessible applications. Imagine a MapQuest application on a cell phone that shows the current location of the device. once people can fix their locations and transmit these coordinates to other devices, a full range of applications is possible. These include location-based services (find the closest automatic teller machine), advertising (get a coupon for a discount at a fast food restaurant around the corner) or social networking (find nearby friends).

The ability of individuals to accurately determine and record locations in the field is also revolutionizing the way geographic data is collected and compiled. Using GPSenabled devices, thousands of amateur users act as citizen sensors that routinely create volumes of volunteered geographic information (vGI). For example, citizens in New Jersey are locating and reporting wetland features. People can use personal navigation systems to send data to vendors about changes in road features and points of interest. For example, openStreetMap has fostered a worldwide phenomenon in which thousands of participants freely form mapping parties to create their own street maps.

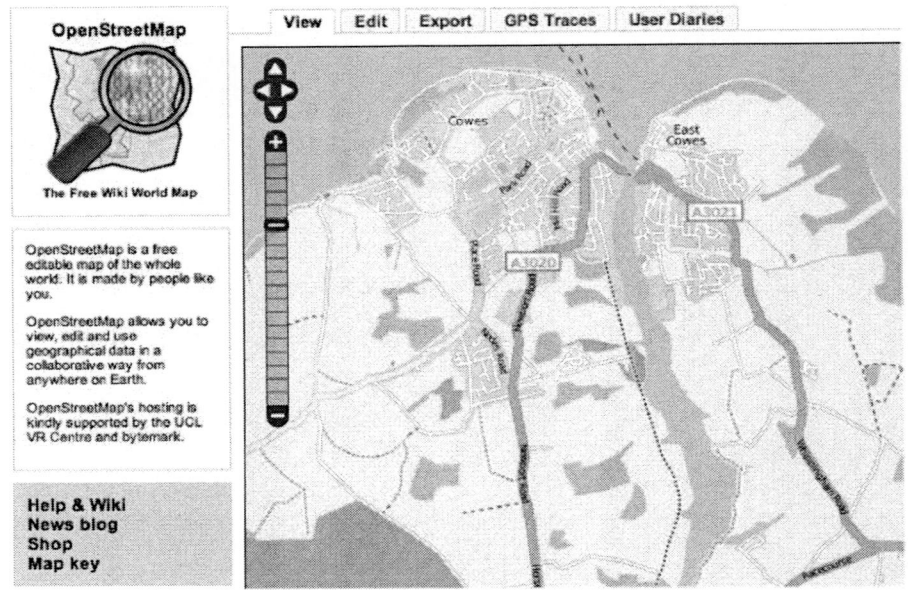

OpenStreetMap's website

Social mapping capabilities are changing long-held constructs of map production and use. In many parts of the world maps have long been hoarded as military intelligence property. In these regions, map data is now being captured in the field by volunteers riding bicycles or walking. organizations such as openStreetMap process this community data to create maps. Some of these maps may be the only available map for an area. The availability of these maps on the Web puts geography in the hands of everyone.

THE EVOLUTION OF GIS: FROM INSTITUTIONS TO VIRTUAL GLOBES

The development of digital mapping software began in earnest in the 1970s with the advent of the first software programs that could convert existing maps into digital data. These early systems ran on large mainframe computers that only existed in large public organizations. In the US, the period was dominated by federal agencies such as the USGS and the Census Bureau that developed their own mapping software to create and maintain digital representations of their existing paper maps. In addition to map generation, these systems were used to conduct inventories of land use and limited integration with other data layers. The Census Bureau developed a system called geocoding to automatically assign coordinates to a street address. These agencies now have employ commercial software for their enterprise-wide geographic information systems (GIS). After a decade, some innovative industries such as timber and utilities, along with a few state agencies and large local governments, were operating their systems on dedicated minicomputers. In a 1983 report the National research Council suggested that the creation of an integrated, nationwide GIS could conceivably manage millions of tax parcels. This foresight was an inkling of GIS's potential for managing vast spatial data infrastructures.

The decade of the 1980s represented a migration of geographic information technology to affordable integrated graphics workstations and client-server environments, which facilitated the sharing of data across a network. This enabled the technology to be adopted by hundreds of midsized organizations and agencies. These organizations often relied on medium-scale digital databases that had been created by federal agencies. These data sources supported applications based on relatively crude scales such as street centerlines and administrative areas and land use. Tremendous inroads were made in the use of multiple layers of data for planning applications, suitability analysis, reapportionment and other census-based data.

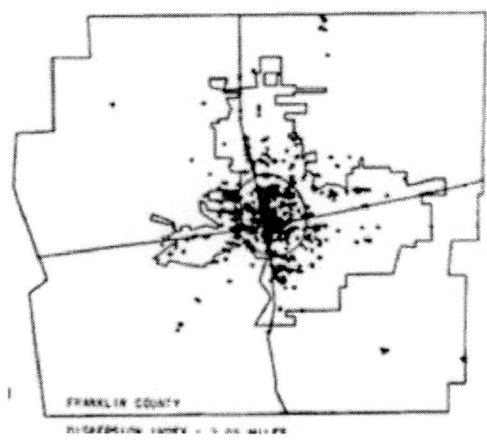

Early GIS

Using commercially available software tools from GIS companies such as ESRI and Intergraph, organizations began to create and maintain extensive geographical databases of corporate and public assets. Most of the analysis consisted of projects that addressed specific issues rather than the daily business activities of an organization. These projects were performed by skilled technicians who knew how to find and use the proper set of software tools and the output was often a printed report with tables and maps. Dramatic advancements were made in tools to manage images and model terrain. Commercial digital image processing tools from companies such as ERDAS could convert aerial photographs into geographic data. At that time, digital photography technology was limited and satellite data was only useful for large-scale reconnaissance of activities such as agricultural production.

By the 1990s, improvements in computer hardware and software provided a watershed for the democratization of computing and GIS software. Agencies migrated their GIS from UNIX to Microsoft Windows operating systems and from specialized workstations to common personal computers. Software was accessed through easy-to-use graphical user interfaces (GUIs). Performance improved as the industry provided faster and cheaper processors, graphics cards and storage systems. These advancements meant that powerful GIS software could be used both by technical "chauffeurs" who created projects and by non-technological professionals such as decision makers, planners, scientists and students. Consequently, GIS was successfully adopted by thousands of local government and business users.

Workstation GIS

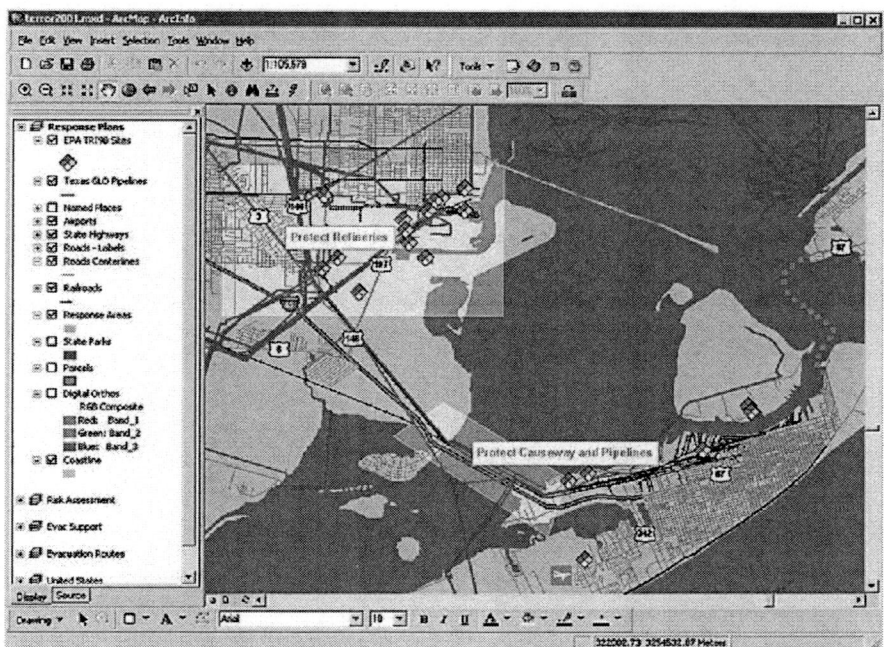

GIS for emergency response

Other events improved the level of common user adoption. Ready and free access to digital versions of Census TIGER files and US Geological Survey topographic quadrangles provided a fundamental base map of the nation that could be added to GIS mapping projects. Universities established teaching labs and helped to train a labor force familiar with geospatial science and applications. By the end of the decade, personal computers were linked to internal networks and the Internet. These advancements allowed for free online Web mapping services that could be easily accessed and used by average citizens. The era of location-based advertising emerged. Commercial GIS software expanded to include hundreds of tools to integrate different kinds of information, process images, perform site analysis, support decisions and generate high-quality cartography.

GIS software could incorporate digital imagery and computer aided design (CAD) and it could generate publication-quality maps. Satellite imagery with 15-meter resolution was also widely available and GPS technology was changing the way surveying and earth measurements were performed. It should also be noted that during this period the traditional paper map-based National Mapping Program operated by the US Geological Survey was all but eliminated. This topographic map series had provided the blueprint for the development of much of the nation and provided critical information for development of our natural resources. It can be argued that the reduction of this program has greatly diminished the federal role as authoritative source of geospatial information.

In the 21st century there has been steady increase in the number of commercial desktop software users who are able to create, maintain and analyze an extraordinary range of geographic information. Moreover, the emergence of the complementary, new generation of Web-based GIS has made it often irrelevant as to whether an application is running on a desktop or across the Internet. This new computing environment has essentially enabled the integration of a geographic perspective within almost every possible information domain.

GIS professionals rely on desktop software to develop tools, and they use the Internet to deploy them to a vast array of consumers. These people are producing a seemingly limitless range of applications such as realistic three-dimensional visualizations and tools for integrating geospatial technologies with spreadsheets and other standard databases. This transparency has been fostered by open systems and open data standards that result in enterprise environments, which provide services on the open Web. From a technical viewpoint, it is important these applications be built with reusable software components that have been developed with object-oriented and scripting languages.

Many traditional barriers to participation in the geospatial data environment have disappeared. Rather than maintaining large staffs and infrastructure, organizations can now build entire applications without purchasing or storing any data or large toolkits. These capabilities have opened the door for GIS professionals to serve an exciting new market with customized applications, support for executive decision-making, and simplified tools that meet the needs of the task-specific or casual user.

The ability of networks to link to remote servers has empowered a new breed of knowledge experts and mobile and location based services, as well as traditional GIS professionals. The creation of huge server farms spread across extensive broadband networks has eliminated the need for users to acquire, download and store massive volumes of data and imagery. often, images and pre-rendered maps are accessed for geographical context and the spatial search or analysis is conducted on a remote server.

The Changing Geospatial Landscape 63

GIS on mobile devices

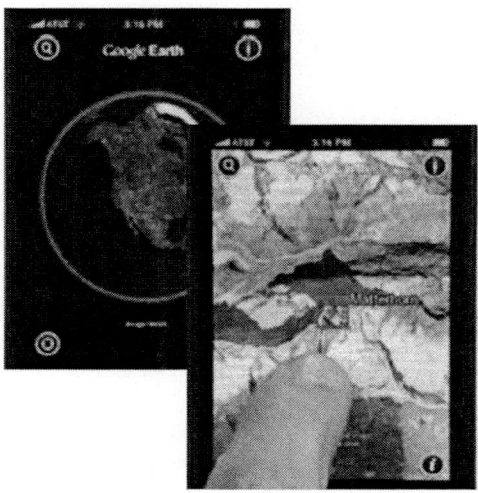

Google Earth application on iPhone

Applications that once were performed by an application specialist on a desktop are now pushed to a server and quickly and seamlessly accessed by a host of users on a wide range of devices. This has enabled handheld devices to become powerful tools. Thousands of GIS professionals employed by the Census Bureau and hundreds of other organizations can go into the field with an inexpensive handheld device to capture new attributes or update existing ones and wirelessly transmit this data back to the office. Similarly, average citizens can now access Google earth on their iPhones to determine their current location or to find a good restaurant.

Some experts suggest that emphasis should shift toward the technical and institutional infrastructure to support the distribution of geographic information throughout society. These spatial data infrastructures (SDI) are frameworks that incorporate technologies, policies, standards and human resources to store, process and distribute vast amounts of data across many organizations and among governments. In the United States, the development of SDIs began in 1994 when President Clinton issued an executive order to create the National Spatial Data Infrastructure (NSDI) and form the Federal Geographic Data Committee (FGDC). This mandate validated the essential role geographic information plays in modern society. The order drove systems to be better coordinated and less redundant. less emphasis was placed on products and more attention was given to processes, knowledge infrastructure, capacity building, communication and coordination. In an Internet-based world, value reaches beyond simply sharing data, and extends to judging data quality and to determining data fitness for consumption. With this, came the necessity to document data in a manner similar to documenting a library's book catalog. Database quality and content took on a different meaning as public agencies published their data via Web browser-based applications that allowed average citizens to query and view detailed information about their property.

Much emphasis in the 21st century has been placed on providing accurate data to support decision-making. In the public and commercial arena, these decisions are diverse. organizations want to know how to pursue an enemy on a battlefield; what are the best land use alternatives for combating global warming; where should police be assigned to reduce crime; what areas are at risk for West Nile virus; what is the best site to build new schools; or what are the route logistics for efficient delivery truck fleet management. At a personal level, people want to know how to get to a party, where to vote, what neighborhood is a good location to buy a house, where to find their friends, and how will an ambulance find them when they call 911.

Today's citizens, taxpayers, and homeowners have an entirely different set of geographic information needs and expectations than people did thirty, twenty or even eight years ago. They want to access geographic information from home through powerful, inexpensive personal computers by means of broadband networks. People accustomed to social Internet structures are as interested in publishing as they are in consuming information. They will readily participate in Facebook's "what are you doing now" dialog. Today's generation of Internet users are often armed with their personal navigation system, are repeat consumers of Google earth data, and expect easy-to-use applications such as seeing their homes and relational values. They flock to sites such as Zillow.com and Cyberhomes. com to view the value of their property and observe the trends in their neighborhoods. This cyberspace generation has high expectations of geographic technologies. They expect to link to their local assessor's records. They expect detailed, recent aerial photography, and, even better, with bird's-eye views at four different oblique angles. In reaction to these demands, local governments are incorporating GIS into their enterprise-wide IT environments. Waukesha, Wisconsin, for instance, reports that scores of business decisions relating to everything from E91 1 to school zoning are driven from a parcel-based GIS because it is the expected norm.

Development approaches change dramatically when designing systems that meet the needs of users who are homeowners and taxpayers. Governor o'Malley of Maryland recently stated,

> ...I'd like you to consider the answer to this question – why is it that virtually any display of GIS technology quickly inspires someone one to ask the timeless question, "... Can you show me my house?..." Through the power of mapping, we were able to create our city's [Baltimore] first-ever complete inventory of housing stock including the ownership information that could be used and accessed by mangers of boarding and cleaning crews, by those responsible for policing, those responsible for inspections, those responsible for filing the lien on the property after cleaning, those in the city's housing department responsible for clearing title, and taking title, and those responsible for disposing of title so the property could be redeveloped and returned to the tax rolls.

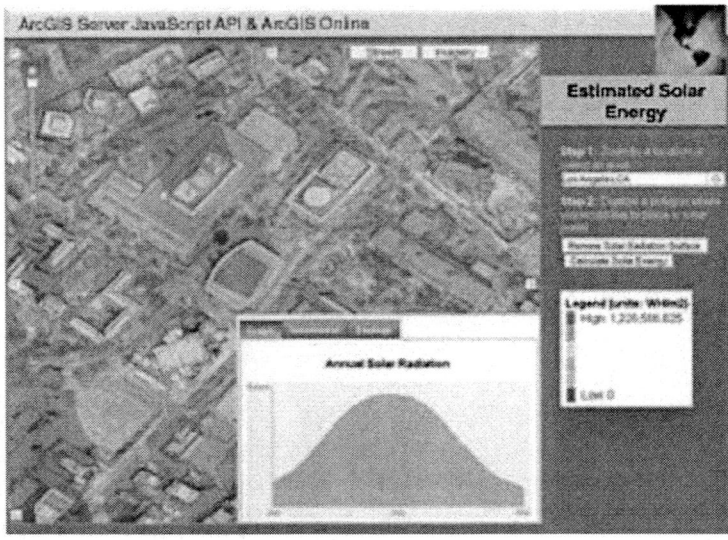

GIS application calculates solar energy potential in Boston

To meet the expectations of these new users that include citizens, public employees, and real estate-associated professionals, a unified approach is required. Property lines must be accurately depicted, images must display fine details (new additions and renovations), and 3D terrain models must model the flow of water through a neighborhood. These needs can only be met by investments in new data and geographic information tools that integrate vast amounts of very high-resolution data that is often measured in terabytes.

THE EVOLUTION OF GIS: THE NEW WHITE BOARD

The previous discussion suggests that the evolution of geographic information technology into mainstream consumer applications had its origins in investments and innovations made by the federal government. At the beginning of this transformation, a single individual or sometimes a small group of scientists could post information into a single computer and see limited results. But barriers still existed for that group to publish results to a wider audience. Now, current IT infrastructure encompasses federated, Web-based, and private-sector approaches. This changing landscape affects and is affected by the federal government as well as multi-collaborative stakeholders.

Significant advances in technology have changed the relative roles of different stakeholders as well as the markets' environment. It is hard to ignore the importance of the recognition by Microsoft, Apple and Google of the business case for location-based searches and applications in changing a field that was once dominated by the public sector GIS professionals. Now the resulting data and software generated by the dedicated GIS community can be leveraged by the exploding group of casual GIS consumers.

The earth is a huge study area. It can be divided into pieces of various sizes and studied at macro or micro scales. For some applications, such as tracking hurricanes, scientists can rely on relatively coarse-grained information but need it updated in real time. Conversely, a civil engineer may require centimeter-level precision when constructing a new bridge. The history of geographic information applications has been one of making trade-offs. A person could either study large areas at crude levels of detail or small areas in fine detail. As we approach the end of the first decade of the 21st century, these trade-offs no longer apply. Perhaps no application exemplifies the success of this better than Google earth. When released in June 2005, Google earth represented a paradigm shift that shook many of our established perceptions about geospatial data. It offered multi-scale, full earth visualization that was free, easy to use and provided a dynamic sense of travel. even though several examples of large-scale, robust geospatial databases existed, none could match Google earth's ability to fly virtually to any place on earth and visualize information at fine detail. Because it is free and easy to use, its success has skyrocketed over the past three years. Content from scores of sources (National Geographic, New york Times, youTube etc.) has been geographically tagged.

A recent article, "Armchair Archaeology" in The economist, describes how Google earth is changing the way archaeologists "make discoveries, develop theories and plan expeditions." The archeologist states, "Google earth gives you free access to imagery that would otherwise cost a fortune and require specialist training to make use of." A conservative estimate of the number of Google earth users is more than 100 million. The net result is that in just three decades, the number of geographic data users has grown from tens of thousands, to a few hundred thousand and then almost instantaneously jumped to hundreds of millions. Its impact has been widely documented in the popular press by experts such as James Fallows of Atlantic Monthly who considers Google earth to be the fourth major innovation in popular computing (along with text editing, the Internet, and the Web). It is so mainstream that it has been the subject of New yorker cartoons and Google earth for Dummies is now a popular reference. More importantly, Google earth has actually become a common platform for hosting and sharing geographically referenced content of all kinds. In many ways, the mapping service has emerged as the new geographic whiteboard, with hundreds of millions of users posting, consuming and comparing data collaboratively on a common earth study area. This simple-to-use visualization tool is valuable complement to the professional GIS tools that continue to be used to develop content, execute spatial analysis and perform modeling to support businesses and governments across the country. The value of spatial data and visualization is being realized simultaneously by casual users and professionals.

CONSIDERATIONS AMIDST THE SEA CHANGE

The demonstrated public appetite for spatial information will require a substantial, educated GIS workforce to meet the demand. The Geospatial Information and Technology Association reported that the geospatial sector has steadily increased by 35% a year, with the commercial side growing at an incredible rate of 100% annually. The US Department of labor predicted that geospatial was one of the three technology areas that would create the most jobs in the coming decade and importantly these are high tech and good paying jobs. All of these changes in terms of users and expectations have turned the traditional governmental and commercial relationships upside down. Most noteworthy has been the dramatic shift of the federal government from being the primary provider of geographic data to that of a major consumer. With a few exceptions for administrative regulations such as the decennial census and flood plain boundaries, local governments create their own data from in-house resources or commercial providers. In times of emergency, the federal government must acquire the most detailed and current data from these local governments. With companies such as Microsoft and Google as customers, commercial data providers – Navteq, TeleAtlas, Pitney Bowes, First American – are doing a brisk business.

Demand for high-resolution imagery from both aircraft and satellite platforms has increased. The recent launch of GeoEye-1 provides a glimpse of the new relationships between private and public organizations. This satellite-based camera is capable of collecting black-and-white images with a 0.41-meter ground resolution and 1.65-meter color images. The major customers for these images are the National Geospatial Intelligence Agency and Google. Aerial photography companies are competing to put fleets of aircraft in the air.

Google Earth

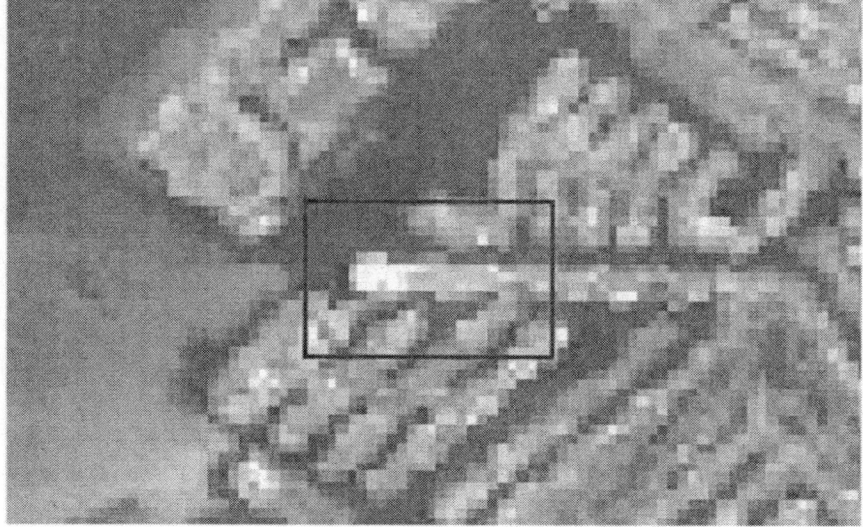

High-resolution imagery: Landsat 30-meter image (above) compared with Digital Globe/Quickbird 1-meter image (right)

Bird's Eye View of Chicago on Microsoft Virtual Earth

These aircraft are equipped with sophisticated digital cameras that can capture huge quantities of geographically registered images. The imagery capabilities allow for billions of pixels, each covering an area as small as a few inches. Pictometry offers data that provides four-inch pixel images from five viewpoints. Applications such as Microsoft's popular Birds Eye View produce images that have added a whole new perspective to house hunting. Airborne lasers that collect detailed elevation data (Light Detection and ranging or LiDAr), provide three-dimensional geographic visualization. These lasers have been characterized as the equivalent of sending thousands of surveyors into the field to collect X, Y and Z coordinates. As a result it is possible to improve the accuracy of flood plain determination and the potential impact of sea level raise in coastal areas.

CHANGING ROLES REQUIRE NEW PARTNERSHIPS AND POLICIES

In recent decades a shift has occurred within the data production community from government to private sector providers. This shift has been encouraged by Congress and the executive branch. A good example of this phenomenon has been the evolution of U.S. commercial remote sensing space policy. The policy has sustained and enhanced the domestic remote sensing industry while advancing and protecting national security and foreign policy interests. The increased involvement of private sector data providers has been fostered by professional organizations and associations such as American Society for Photogrammetry and remote Sensing, American Congress on Surveying and Mapping, and Management Association for Private Photogrammetric Surveyors that support public-private partnerships. (A complete list of the members of COGO, the Coalition of Geospatial Organizations, is on page 13 of this document.) Consequently, government's role has shifted from data producer to coordinator, partnership facilitator, and manager. This, in turn, has resulted in significant growth in the number, size, capacity and capabilities of the US private geospatial community. This community is the most robust in the world, engaged in serving the domestic market and is a significant exporter of services, data and technology to serve a growing global market.

The relative shifts in data production from the federal government to the private sector and state and local government call for new forms of partnership. Furthermore, the hodgepodge of existing data sharing agreements are stifling productivity and are a serious impediment to use even in times of emergency. There is an urgent need to reexamine the relationships between data providers and users to establish a fair and equitable geospatial data marketplace that serves the full range of applications. When the federal government was the primary data provider, regulations required data to be placed in the public domain. This policy jump-started a new marketplace and led to the adoption of GIS capabilities across public and commercial sectors. However, these arrangements are very different when data assets are controlled by private companies or local governments.

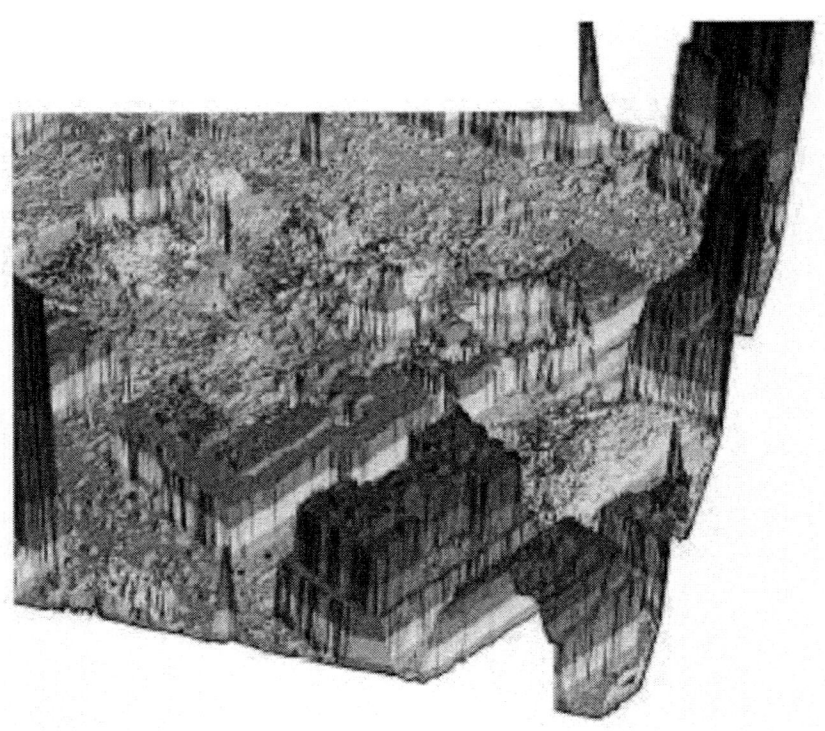

LiDAR image of Ground Zero (Fugro EarthData for the State of New York)

Insistence on database ownership is an expensive policy. When the Census Bureau was updating the street networks to prepare for the 2010 Census, it could not take advantage of the existing commercial data from Navteq or TeleAtlas; therefore, the government spent hundreds of millions of dollars to develop a duplicate version of street centerlines. The Bureau which pioneered the field has attempted to assemble street network data collected from more than 4,000 local governments. They found that data often did not exist, was incompatible or was unavailable because of local licensing policies. Similarly, the federal government's need for tax parcel information has proven a costly venture. Critical information about the use, value and ownership of property is needed by FeMA, the Forest Service, and HUD, for emergency preparedness or response at times of hurricanes or wildfires – or even to monitor the current foreclosure problems. Unfortunately, no arrangements have been made for the federal government to acquire the detailed property-related data that it needs to make responsive decisions. Ironically, private companies such as the online real estate service Zillow are often better prepared than the federal government to support these critical decisions.

The dramatic shift in the relative roles of the federal, state and local governments has been monitored by several institutions and advocacy groups. For example, the National research Council, which oversees the Mapping Science Committee, has conducted numerous studies identifying trends and recommending changes that would improve efficiency and coordination of geographic information. State governments have also emerged as an increasingly important source of intermediate level geographic information coordination.

As early as 1989, several state GIS managers convened as the National States Geographic Information Council (NSGIC) to establish a forum for coordinating GIS projects and government investments. This group provided an early indication of the existence of duplicative efforts and the potential of redundant government activities. NSGIC is one of the most active proponents of spatial data infrastructure projects and almost every state now has a state GIS coordinator. NSGIC has an active agenda and is working closely with the FGDC for new initiatives. One of these initiatives is Imagery for the Nation. It is a model for new partnerships in which the federal government provides partial funding to acquire high-resolution digital imagery collected by commercial data providers, with the option for state and local governments to "buy-up" for higher-resolution data. This data will be placed in the public domain and will be freely available to all sources including commercial entities such as Google and Microsoft who will use this data to fuel their product and service offerings to the marketplace.

Nearly all the data, technology and applications we see today can be traced to innovative policies and government practices of the past. As such we require similar innovative policies now to keep pace with this remarkable sea change. Government-based geographic information providers can no longer think of themselves as a players outside of or immune from the community of private sector, state, local or even public stakeholders. In many cases these stakeholders have embraced technology and processes which have rapidly outpaced anything the federal government can provide. At a minimum, what is needed is a commitment to improved spatial data, recognition of the place of multiple stakeholders in this brave new world, and coordinated investment.

Although phenomena such as the Zillow Website's millions of hits, cars equipped with navigational devices, and phones embedded with location-based services for locating friends are fascinating, the greatest value of the spatial data infrastructure still lies in illuminating complex policy problems. If we as a country are sincere about resolving universal concerns such as global warming, sea level rise, and affordable health care, the Federal government needs to adopt innovative policies supporting a dynamic and robust spatial data infrastructure, an initiative that was promised more than 15 years ago. The members of the National Geospatial Advisory Committee look forward to working with the Obama Administration and the geospatial community in formulating recommendations on the adoption and or revision of spatial data policies and programs that can empower better decision-making through geography at all levels of government and in private enterprise.

The founding members of the Coalition of Geospatial organizations: American Congress on Surveying and Mapping (ACSM); American Society of Photogrammetry and remote Sensing (ASPrS); Association of American Geographers (AAG); Cartography and Geographic Information Society (CAGIS); Geospatial Information Technology Association (GITA); GIS Certification Institute (GISCI); International Association of Assessing officers (IAAo); Management Association for Private Photogrammetric Surveyors (MAPPS); National States Geographic Information Council (NSGIC); University Consortium for Geographic Information Science (UCGIS); Urban and regional Information Systems Association (UrISA). Founding advisory organizations are: National Association of Counties (NACo); National emergency Number Association (NeNA); Western Governors Association (WGA); American Planning Association (APA)

This report was prepared by a subcommittee of the National Geospatial Advisory Committee. This report was approved at the october 2008 meeting of the NGAC. Subcommittee members: **Dr. David J. Cowen**, Distinguished Professor Emeritus, University of South Carolina (Chair); **Dr. Sean Ahearn**, Professor of Geography, Director, Center for Advanced Research of Spatial Information (CARSI), Hunter College – CUNy; **Mr. Michael Byrne**, GIS Architect, State of California office of Statewide Health Planning and Development. The subcommittee would like to thank ESRI and the National Geographic Society for their assistance in the preparation of this report.

For information about the National Geospatial Advisory Committee, please visit **http://www.fgdc.gov/ngac**

Cover illustration: Montage of Mount St. Helens (DigitalGlobe via Google earth) and a GIS-produced map of Los Angeles, CA (courtesy GreenInfo Network).

MEMBERS OF THE NATIONAL GEOSPATIAL ADVISORY COMMITTEE

Ms. Anne Hale Miglarese (NGAC Chair), Principal, Booz Allen Hamilton, McLean, VA

Mr. Steven P. Wallach (NGAC Vice-Chair), Technical Executive, U.S. National Geospatial-Intelligence Agency, Bethesda, MD

Dr. Sean Ahearn, Center for Analysis and Research of Spatial Information, Hunter College – City University of New York, New York, NY

Dr. Timothy M. Bull Bennett, TCU Science Coordinator, North Dakota Association of Tribal Colleges; Executive Director, Nativeview Inc., Bismarck, ND

Mr. Michael Byrne, Geographic Information Architect, State of California office of Statewide Health Planning and Development, Sacramento, CA

Mr. Allen Carroll, Chief Cartographer, National Geographic Society, Washington, DC

Mr. **Richard B. Clark**, Chief Information officer, State of Montana, Helena, MT

Dr. **David J. Cowen**, Department of Geography, University of South Carolina, Columbia, SC

Mr. **Jack Dangermond**, President, Environmental Systems Research Institute, Redlands, CA

Mr. **Donald G. Dittmar**, Land Information System Manager, Waukesha County Department of Parks and Land Use, Waukesha, WI

Mr. **Dennis B. Goreham**, Manager, Automated Geographic Reference Center, State of Utah, Salt Lake City, UT

Ms. **Kass Green**, President, The Alta vista Company, Berkeley, CA

Hon. **Randy Johnson**, County Commission Chair, Hennepin County, Minneapolis, MN

Mr. **Randall L. Johnson**, MetroGIS Staff Coordinator, Metropolitan Council, St. Paul, MN

Dr. **Jerry J. Johnston**, Geographic Information officer, U.S. Environmental Protection Agency, Washington, DC

Mr. **Barney Krucoff**, GIS Director, District of Columbia, Washington, DC

Hon. **Timothy Loewenstein**, County Supervisor, Buffalo County, Kearney, NE

Dr. **David F. Maune**, Senior Associate, Dewberry, Fairfax, VA

Mr. **Charles Mondello**, Senior vice President, Corporate Development, Pictometry International, Rochester, NY

Mr. **Zsolt Nagy**, Program Manager, Center for Geographic Information & Analysis, State of North Carolina, Raleigh, NC

Ms. **Kimberly T. Nelson**, Executive Director for eGovernment, Microsoft Corporation, Washington, DC

Mr. **Matthew O'Connell**, President and Chief Executive officer, GeoEye, Dulles, VA

Mr. **John M. Palatiello**, Executive Director, Management Association for Private Photogrammetric Surveyors, Reston, VA

Dr. **Jay Parrish**, Director and State Geologist Pennsylvania Bureau of Topographic and Geologic Survey, Harrisburg, PA

Mr. **G. Michael Ritchie**, President and Chief Executive officer, Photo Science, Lexington, KY

Mr. **David Schell**, Chairman and Chief Executive officer, open Geospatial Consortium, Wayland, MA

Mr. Eugene A. Schiller, Deputy Executive Director, Division of Management Services, Southwest Florida Water Management District, Brooksville, FL

Dr. Christopher Tucker, Senior vice President, Americas & National Programs, Erdas, Inc., Alexandria, VA

Mr. Ivan DeLoatch, Staff Director, Federal Geographic Data Committee (NGAC Designated Federal officer)

David J. Cowen is a Distinguished Professor Emeritus at the University of South Carolina. During his career at the University of South Carolina he was chair of the Department of Geography, Director of the Liberal Arts Computing Lab, co-director of the Center for GIS and remote Sensing, and a Carolina Distinguished Professor. He is currently a member of the NRC Board on earth Sciences and resources, the Vice President of the Geographic Information Systems Certification Institute and a National Associate of the National Academy of Sciences. Between 2000 and 2006 he chaired the Mapping Science Committee of the National research Council and recently chaired the NRC Study Committee "Land Parcel Databases: A National Vision". He is the 2005 recipient of the ESRI lifetime Achievement Award in GIS. Since 1967 his research and teaching interests have focused on the development and implementation of geographic information systems in a wide range of settings.

Chapter 4

THE NATIONAL GEOSPATIAL TECHNICAL OPERATIONS CENTER[*]

United States Geological Survey

"Providing critical geospatial information to the Nation"

The United States Geological Survey (USGS) National Geospatial Technical Operations Center (NGTOC) provides geospatial technical expertise in support of the National Geospatial Program in its development of *The National Map*, National Atlas of the United States®, and implementation of key components of the National Spatial Data Infrastructure (NSDI).

MISSION STATEMENT

The NGTOC provides essential support for the acquisition and management of trusted geospatial data, products, and services through nationally recognized geospatial technical expertise and customer service for the USGS and the Nation.

---

[*] This is an edited, reformatted and augmented version of United States Geological Survey, dated March 2009.

MAJOR PRODUCTS AND SERVICES

The National Map

The National Map, a significant component of the National Spatial Data Infrastructure (http://www.fgdc.gov/nsdi/nsdi.html), is a collaborative effort by the USGS and partners to improve the utility, currency, and availability of national topographic information. The NGTOC plays a key role in this effort by enhancing the usefulness of national geospatial products and services, and by acquiring new geospatial data, assessing the data for accuracy and quality, integrating it into nation-wide datasets, and improving public access to that information through online viewing and data download.

Essential services supporting these activities include research and development of national geospatial data standards, data models, and products that keep pace with state-of-the-art technology and expanding user requirements. The NGTOC maintains a staff of geospatial experts who serve all levels of government in acquiring high- quality geospatial products and services from world-class contractors. The NGTOC provides training and technical support to partners for advancement of the NSDI and to geospatial data users at all levels of government, as well as the general public.

The National Map Products

- USGS Topographic Maps: The NGTOC is developing the next generation of USGS topographic map products. These products are available through the USGS Store: http:// store.usgs.gov
- Orthoimagery: The NGTOC acquires and quality checks orthoimagery before its integration into the national seamless orthoimagery dataset. USGS orthoimagery and information may be accessed at: http://gisdata.usgs.net/website/ Orthoimagery/
- Elevation: The NGTOC acquires, processes, and quality checks terrain elevation data to prepare it for integration into the seamless National Elevation Dataset found at: http://ned.usgs.gov
- Hydrography: The NGTOC provides National Hydrography Data (NHD) database management, data model enhancement, and data improvement through internal operations and training and support to

data stewardship partners. The National Hydrography Dataset is available at: http://nhd.usgs.gov

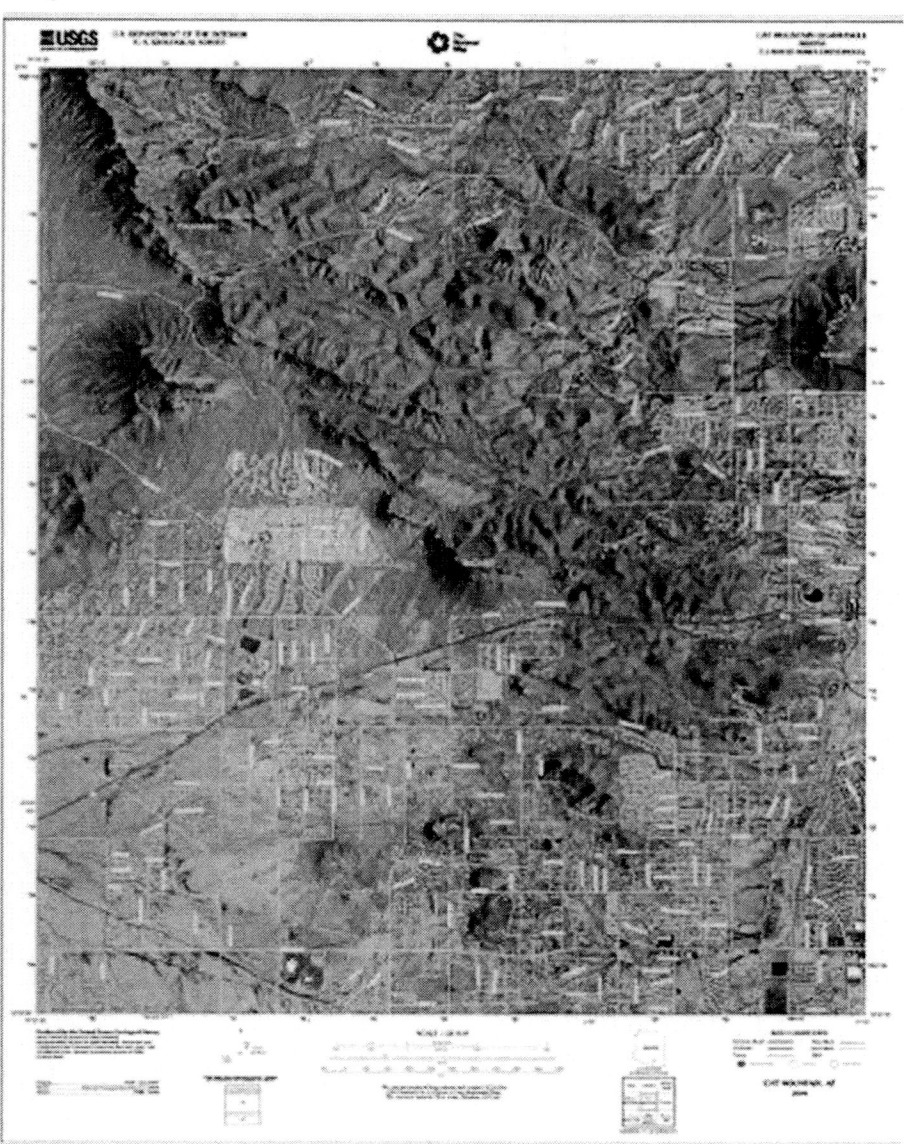

Quadrangle map of Cat Mountain, Arizona.

- Geographic Names: The NGTOC provides support to the U.S. Board on Geographic Names and works with private sector and Federal, state, and local partners to develop and maintain information about geographic names for the Nation. Information about these activities can be found at: http://geonames.usgs.gov/domestic
- Boundaries, Transportation, and Structures: The NGTOC works with numerous partners to acquire and maintain a nationally consistent database of boundaries, transportation, and man-made structures information. More information about these datasets can be found at: http://bpgeo.cr.usgs.gov/

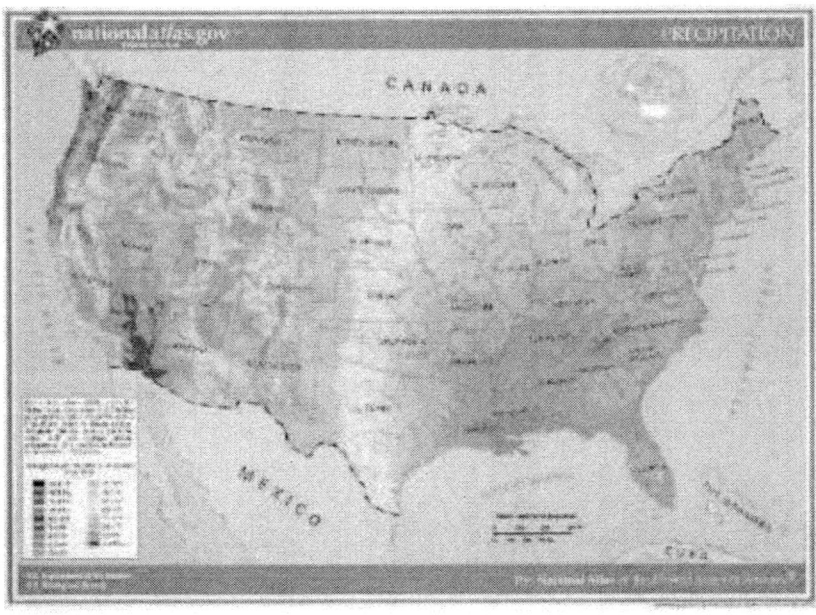

National Atlas: U.S. Precipitation

National Atlas of the United States®

The National Atlas of the United States® provides a comprehensive, map-like view into the enormous wealth of geospatial and geostatistical data collected for the United States. It is designed to enhance our geographic knowledge and understanding of the physical and cultural geography of the U.S. NGTOC staff compile and integrate small-scale (1:1,000,000-scale)

National Atlas geographic information system (GIS) data and support the data integration and documentation needs of the program's Federal partners. Experienced NGTOC cartographers and technicians are harmonizing these mapping frameworks with similar cartographic data from Canada and Mexico. Their work results in valuable contributions to the Atlas of North America (http://www.cec.org/naatlas/) and the Global Map (http://www.globalmap.org/english/index.html). NGTOC staff members provide quick and responsive customer support for all eight National Atlas products and services. National Atlas education and outreach efforts also are carried out through the NGTOC. In addition, the Center hosts the online National Atlas ensuring reliable, uninterrupted service from nationalatlas.gov (http://nationalatlas.gov/)

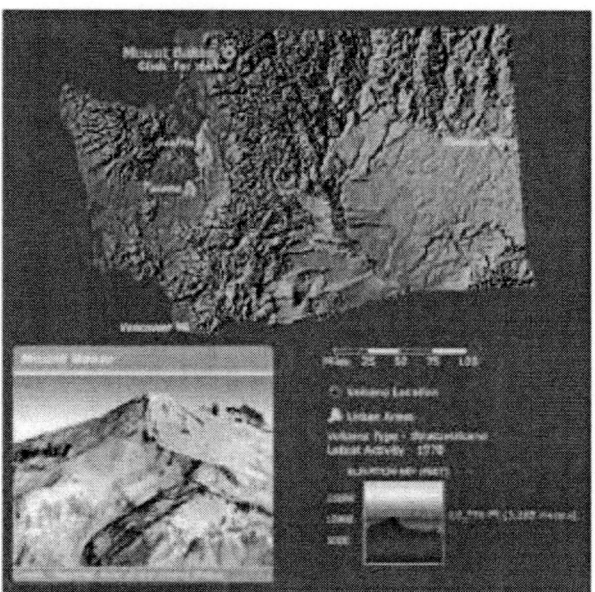

Potentially Active Volcanoes in Washington

In: Geospatial Information and GIS: Background... ISBN: 978-1-61761-432-3
Editor: Sean C. Dallon © 2011 Nova Science Publishers, Inc.

Chapter 5

WEB-BASED GEOSPATIAL TOOLS TO ADDRESS HAZARD MITIGATION, NATURAL RESOURCE MANAGEMENT AND OTHER SOCIETAL ISSUES[*]

United States Geological Survey

Federal, State, and local government agencies in the United States face a broad range of issues on a daily basis. Among these are natural hazard mitigation, homeland security, emergency response, economic and community development, water supply, and health and safety services. The U.S. Geological Survey (USGS) helps decision makers address these issues by providing natural hazard assessments, information on energy, mineral, water and biological resources, maps, and other geospatial information.

Increasingly, decision makers at all levels are challenged not by the lack of information, but by the absence of effective tools to synthesize the large volume of data available, and to utilize the data to frame policy options in a straightforward and understandable manner. While geographic information system (GIS) technology has been widely applied to this end, systems with the necessary analytical power have been usable only by trained operators. The

---

[*] This is an edited, reformatted and augmented version of United States Geological Survey, dated June 2009.

USGS is addressing the need for more accessible, manageable data tools by developing a suite of Web-based geospatial applications that will incorporate USGS and cooperating partner data into the decision making process for a variety of critical issues. Examples of Web-based geospatial tools being used to address societal issues follow.

MAPPING, MONITORING, AND ANALYSIS OF ENVIRONMENTAL MERCURY

Mercury in our environment—air, water, soil, and especially our food—poses significant hazards to human health, particularly for developing fetuses and young children. Because of the importance of this issue and the length of time it has been studied, large and complex data sets of mercury concentrations in various media and associated ancillary data have been generated by many Federal, State, Tribal, and local agencies. To facilitate the efficient and effective use of these data in managing and mitigating human and wildlife exposure to mercury, the U.S. Geological Survey and the National Institute of Environmental Health Sciences have developed a Web site for visualizing and studying the distribution of mercury. The Environmental Mercury Mapping, Modeling, and Analysis (EMMMA) Web site is designed to support environmental and health researchers, as well as land and resource managers, by providing useful data, map products, and Web-based tools (Hearn and others, 2006; http://emmma.usgs.gov). Components of EMMMA include:

- Online mapping tools, maps, imagery, and other thematic data used to display and analyze mercury data and to print maps (figure 1). USGS mapping data provide a nationwide geographic reference composed of aerial photographs, satellite imagery, and geospatial data for land cover, elevation, hydrology, transportation, and geographic names. These data provide a useful geographic context for the mercury data displayed on top of them.
- An online model to describe mercury in fish tissue, which standardizes the concentration of mercury in fish to enable normalized comparisons among different species, individuals of different lengths, and samples of different types. The National Descriptive Model for Mercury in Fish Tissue (NDMMFT) is applied to a comprehensive

national compilation of fish-tissue data to detect spatial and temporal trends in mercury concentrations that would otherwise be obscured (Wente, 2004). Modeled data from EMMMA can aid researchers in spotting significant trends in the environment, and also can be used by State agencies to develop more comprehensive and cost-effective fish consumption advisories (figure 2).

THE LAND COVER ANALYSIS TOOL

USGS's National Land Cover Dataset (NLCD) is used widely by scientists and land managers to assess changes in land cover that affect water quality, wildlife habitat, human and animal population distribution, carbon storage related to global climate change, and other important environmental, ecological, and societal examples. The Land Cover Analysis Tool (LCAT) is a Web-based tool kit that allows users to quickly locate, display, and download NLCD data, including the recently developed NLCD Change Product, which displays changes in land cover between 1992 and 2001 (http://lcat.usgs.gov).

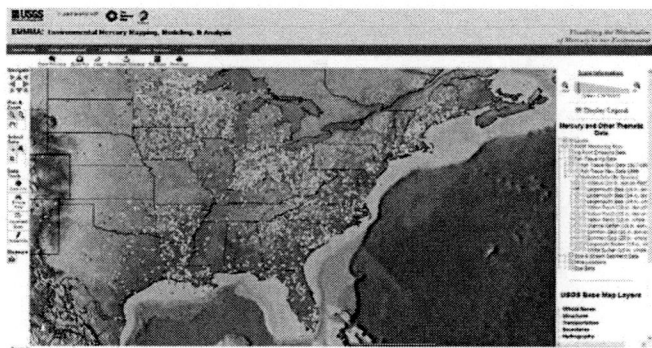

Figure 1. EMMMA's Data Mapper allows users to navigate to and download data from selected locations, and also overlay these data on top of topographic maps, satellite imagery, and aerial photographs.

LCAT allows users to clip, display, and download NLCD data using a variety of polygons, including State, county, city, and watershed boundaries; and uploaded GIS shapefiles, including polygons defined onscreen by the user. Additional features include the ability to display individual land cover classes on top of user-selected maps or imagery and the ability to generate and print reports detailing the land cover composition within selected areas. User-selected change pairs (for example, forest to urban) can als be displayed on top of satellite imagery to provide context for areas of change. An example of the type of analysis provided by LCAT is shown in figure 3.

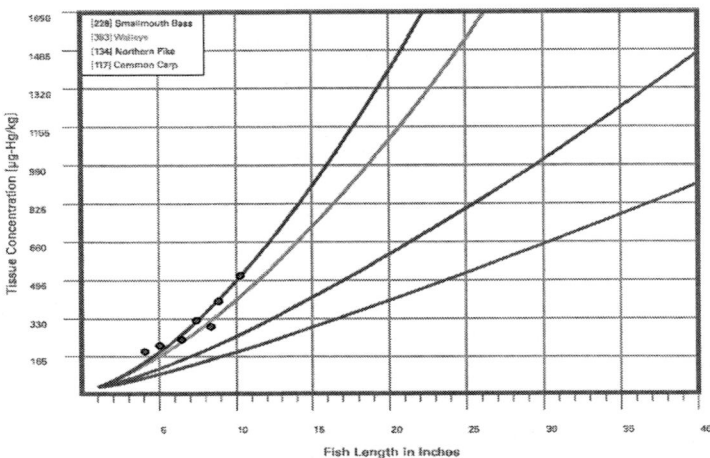

Figure 2. This graph shows a mercury versus fish-length curve fitted by the NDMMFT using mercury data for smallmouth bass (dark blue curve) from a U.S. lake. Using this curve the model can generate curves for other locally occurring species of fish (walleye, northern pike, and carp are shown here) and predict expected mercury content for a given length.

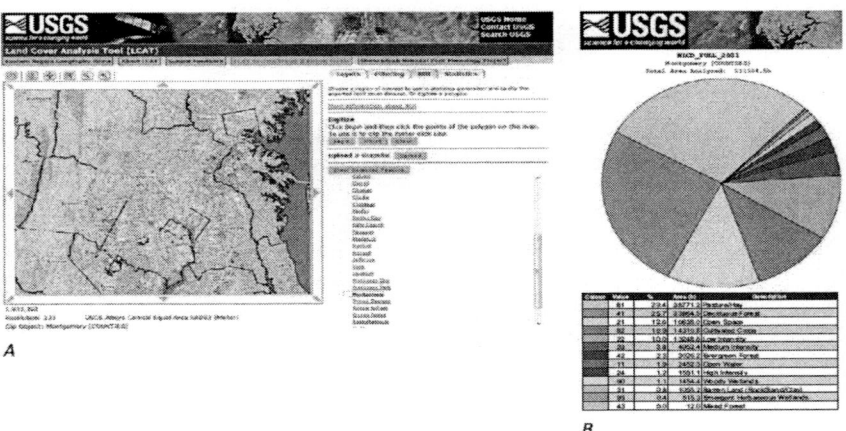

Figure 3. LCAT enables users to rapidly calculate and download various land cover data statistics for a variety of user-defined areas. This figure shows A, the results of calculating the 2001 NLCD for Montgomery County, Md.; and B, a graphical representation in the form of a pie diagram and the data table. LCAT requires no special knowledge or software; this calculation was completed in less than 5 seconds.

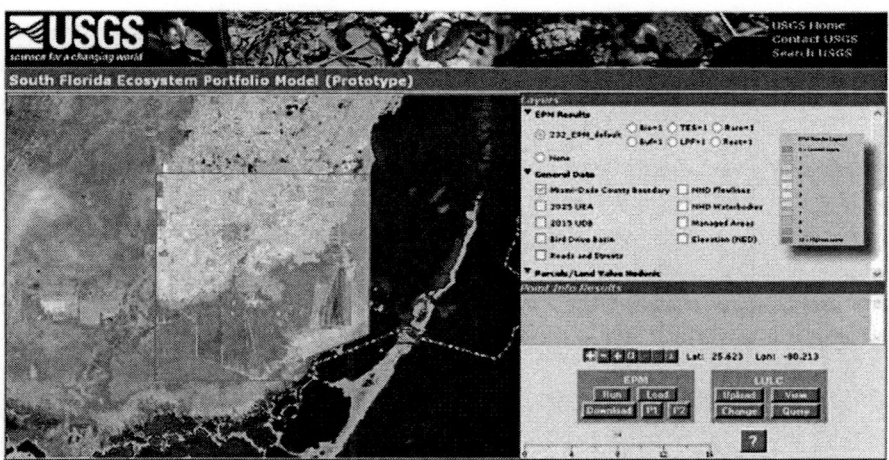

Figure 4. The Ecosystem Portfolio Model allows users to quickly develop suitability maps drawn from elected ecological and other physical data. The red bordered rectangle above shows the extent of an area between the Everglades and Biscayne National Parks that was analyzed for relative suitability. Biodiversity, ecosystem restoration potential, and low landscape fragmentation metrics were chosen as priorities.

THE SOUTH FLORIDA ECOSYSTEM PORTFOLIO MODEL

Intense pressures for development of the land outside of the Urban Development Boundary in Miami-Dade County, Florida, threaten to impact both the Everglades and Biscayne National Parks. In response, the U.S. Geological Survey, in cooperation with the National Park Service and the Wharton School of Economics, has developed a prototype for a Web-enabled geospatial information tool—the South Florida Ecosystem Portfolio Model, or EPM. The EPM allows users to evaluate hypothetical future land use and land cover patterns in terms of habitat availability, indicators of ecological health, water quality criteria, land price changes, and quality of life indicators. The criteria and indicators are chosen to be sensitive to land use and land cover relevant to decision makers and stakeholders, and indicative of future trends. Users may assign weights to these criteria to reflect their views regarding their relative importance. An example of ecological criteria is shown in figure 4, along with the evaluation results. The EPM project uses contributions from conservation ecology, landscape ecology, decision science, real estate economics, ecological economics, urban planning, GIS analysis, and Web technologies. In the next phase, the Web-based EPM will be extended to other areas of South Florida facing development pressure in natural and agricultural areas. This project will contribute to improved public understanding and awareness of the complex ecological, environmental, and socioeconomic issues involved in land use decisions in southeast Florida, and also offers a useful approach for other areas in the United States where there are competing visions for land use.

APPLICATION OF THE LAND USE PORTFOLIO MODEL TO SEISMIC HAZARD MITIGATION IN MEMPHIS, TENNESSEE

The City of Memphis and surrounding Shelby County lie within the New Madrid Seismic Zone, which extends from northeast Arkansas through southeast Missouri and western Tennessee, and western Kentucky to southern Illinois. Historically, Shelby County has been the site of some of the largest earthquakes in North America. Earthquakes with estimated magnitudes between 7.5 and 8.0 occurred in this area between 1811 and 1812 (Gomberg and Schweig, 2003). The City of Memphis and portions of Shelby County have been mapped by the USGS to assess the level of seismic hazard (figure 5; http://earthquake.usgs.gov/regional/ ceus/index.php). Shelby County was chosen as an area to evaluate the Land Use Portfolio Model (LUPM), a modeling, mapping, and risk communication tool that can assist public agencies and communities in understanding and reducing their natural-hazards vulnerability (http://geography.wr.usgs.gov/science/ lupm.html).

The LUPM builds upon financial-portfolio theory, a method for evaluating alternative investment choices based on the estimated distribution of risk and return from different investment options. The LUPM been has developed into an interactive, GIS-based decision support system that stakeholders can use to select locations in which to invest a hazard mitigation budget, evaluate metrics such as the mean and variance of community wealth, and compare and rank policies. The model is unique in that it allows users to think through various levels of risk tolerance and hazard acceptability and to compare the cost effectiveness of alternative policies.

Initial results from LUPM runs demonstrate the effect of earthquake probability estimates, planning horizons, and the cost of mitigation estimates on hazard mitigation investments (Bernknopf and others, 2007). Also, the desktop version of LUPM is being modified to create a Web-based version, which will provide managers and policy makers with a simplified interface to better understand how the model works and how it can benefit the analysis of hazard mitigation policies.

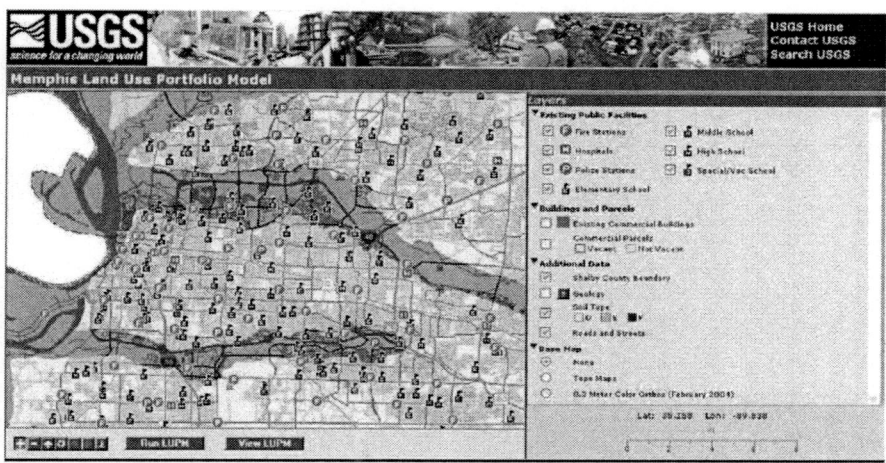

Figure 5. One application of the LUPM in Memphis is to provide a planning tool for municipal authorities responsible for schools, police stations, fire stations, and other public buildings. The LUPM will allow managers to test various scenarios for the mitigation of seismic risk to these assets and evaluate mitigation costs required to lower the risk from specified seismic events.

REFERENCES CITED

[1] Bernknopf, R.L., Hearn, P.P., Wein, A.M. & Strong, David (2007). *The effect of scientific and socioeconomic uncertainty on a natural hazards policy choice*, in International Congress on Modeling and Simulation—MODSIM07, December 2007, Christchurch, New Zealand: Christchurch, Modeling and Simulation Society of Australia and New Zealand Proceedings, 1702–1708.

[2] Gomberg, J. & Schweig, E. (2003). *Earthquake hazard in the heart of the homeland*: U.S. Geological Survey Fact Sheet 131–02, 4 p. (Also available at http://pubs.usgs.gov/fs/fs-131-02/.)

[3] Hearn, P.P., Wente, S.P., Donato, D.I. & Aguinaldo, J. J. (2006). *EMMMA—A Web-based system for environmental mercury mapping, modeling, and analysis*: U.S. Geological Survey Open-File Report

2006–1086, 13 p. (Also available at http://erg.usgs.gov/isb/pubs/ofrs/2006-1086/ofr2006-1086.pdf.)

[4] Wente, S. P. (2004). A statistical model and national data set for partitioning fish-tissue mercury concentration variation between spatiotemporal and sample characteristic effects: U.S. Geological Survey Scientific Investigations Report 2004–5199, 15 p. (Also available at http://pubs.usgs.gov/sir/2004/5199/ pdf/2004-5199.pdf.)

Author Paul P. Hearn,

INDEX

A

accessibility, 29
accuracy, 7, 9, 20, 23, 25, 36, 47, 62, 79, 88
advocacy, 21, 82
aerial photographers, 7
agencies, xi, 2, 4, 7, 9, 13, 17, 19, 21, 22, 24, 26, 27, 30, 34, 45, 49, 51, 52, 54, 62, 65, 66, 72, 93, 94, 95, 99
Alaska, 31, 38
American Recovery and Reinvestment Act, 34
annotation, 12
appetite, 62, 76
appropriations, 51
assessment, 25
assets, 27, 61, 62, 66, 80, 100
authorities, 4, 18, 23, 100

B

barriers, 69, 74
Belgium, 61
bridges, 10, 33
Broadcasting, 36
browser, 72
buildings, 5, 10, 25, 50, 52, 100
Bureau of Land Management, x, 24, 34, 39, 42, 56

C

cables, 26
candidates, 52
capacity building, 72
carbon dioxide, 47
carp, 97
CBS, 36
cell phones, x, 59, 61
Census, ix, 2, 4, 15, 16, 17, 29, 31, 36, 38, 42, 50, 52, 53, 57, 60, 65, 68, 71, 81
City, 13, 14, 15, 16, 38, 85, 99
class, 88
cleaning, 73
climate change, 40, 43, 47
color, v, 15, 77
community, xi, 18, 19, 27, 28, 44, 45, 65, 75, 79, 82, 83, 93, 99
compatibility, 13
compilation, 95
complement, 76
compliance, 32, 49, 54
composition, 96
computing, 67, 69, 76
confidentiality, 54
conservation, 25, 51, 98
consumer demand, ix, 1, 3
consumption, 72, 95

coordination, 3, 4, 5, 7, 17, 27, 28, 29, 30, 31, 33, 40, 45, 48, 49, 53, 55, 56, 72, 82
cost, 2, 4, 21, 26, 27, 31, 38, 50, 54, 62, 75, 95, 99
cost benefit analysis, 38
cost effectiveness, 99
cost saving, 50
covering, 23, 79
criminals, 62
critical infrastructure, 18, 30, 33
crops, 25, 50
currency, 88
cyberspace, 73
cycles, 22

D

data collection, 21, 27
data set, 2, 3, 24, 34, 49, 94, 101
database, 6, 26, 40, 41, 43, 45, 48, 49, 51, 52, 53, 55, 81, 89, 91
database management, 89
datasets, 88, 91
decision makers, xi, 67, 93, 98
delegates, 29, 50
democratization, 67
Department of Agriculture, 31, 43
Department of Defense, 7, 37, 61
Department of Homeland Security, 33, 52
Department of the Interior, 17, 19, 25, 37, 40, 42, 56
deposits, 47
digital cameras, 79
disaster, 8, 23, 29, 40, 44
disclosure, 29, 51
distortions, 7
District of Columbia, 85
drawing, 34

E

early warning, 45
ecology, 98
economic development, 55
economic losses, 13
economy, 18

ecosystem, 17, 48, 98
ecosystem restoration, 17, 98
E-Government Act, 28, 40, 48, 51
emergency management, 29, 44, 54
emergency preparedness, 28, 42, 81
emergency response, xi, 8, 23, 40, 43, 46, 68, 93
EMMMA, 94, 95, 96, 100
enforcement, 28, 33, 49
entrepreneurs, 62
Environmental Mercury Mapping, Modeling, and Analysis, 94
Environmental Protection Agency, 20, 85
evacuation, 11
Everglades, 98
Executive Office of the President, 28, 41, 49, 51
Executive Order, 2, 4, 18, 33, 40, 48
expertise, x, 87, 88
exploration, 47
exporter, 80
exposure, 94

F

Facebook, 72
Farm Bill, 29, 50, 51
farms, 69
fast food, 64
federal funds, 19, 20
Federal Geographic Data Committee, x, 2, 4, 19, 20, 39, 42, 72
federal programs, 53
FEMA, 9, 23, 25, 28, 34, 36, 37, 42
FGDC, 2, 4, 18, 19, 20, 21, 25, 26, 27, 28, 29, 31, 33, 34, 40, 42, 43, 44, 45, 46, 48, 49, 50, 52, 57, 72, 82
fish, 62, 95, 97, 101
Fish and Wildlife Service, 24
fitness, 72
flood hazards, 23
foreclosure, 15, 16, 28, 40, 42, 43, 45, 82
foreign policy, 80
forest ecosystem, 24
fraud, 23, 54
funding, 23, 30, 32, 33, 37, 53, 56, 82

Index

G

General Accounting Office, 36, 37
Geographic Information System, ix, 1, 6
geography, 65, 83, 91, 99
GIS, i, iii, v, vii, ix, 1, 3, 5, 6, 7, 8, 9, 10, 11, 13, 14, 16, 17, 18, 22, 23, 24, 27, 30, 31, 32, 33, 34, 35, 36, 38, 41, 45, 46, 56, 57, 58, 65, 66, 67, 68, 69, 70, 71, 73, 74, 75, 76, 80, 82, 83, 84, 85, 86, 92, 94, 96, 98, 99
global climate change, 95
Google Earth, 3, 71
governance, 32
GPS, ix, x, 1, 7, 35, 59, 60, 61, 62, 64, 68
graph, 97
Great Lakes, 26
greenhouse gas emissions, 48
guidelines, 18, 20

H

habitats, 26
handheld devices, 71
harvesting, 55
Hawaii, 31, 38
hazards, 94, 99
high winds, 6
home ownership, 44
homeland security, xi, 52, 93
homeowners, 72, 73
host, x, 59, 71
hot spots, 45
housing, 8, 44, 45, 52, 73
human resources, 18, 71
Hunter, 84, 85
hunting, 79
Hurricane Katrina, 8, 23
hurricanes, ix, x, 2, 17, 23, 28, 42, 43, 44, 75, 82

I

image, 7, 12, 66, 78, 81
imagery, 21, 31, 38, 68, 70, 75, 77, 78, 82, 94, 96
images, 7, 8, 11, 66, 68, 70, 74, 77
inefficiency, 30
information technology, x, 59, 66, 74
infrastructure, 18, 30, 34, 43, 75, 82, 83
inspections, 73
institutional infrastructure, 71
integration, ix, 1, 35, 62, 65, 69, 88, 89, 92
intelligence, 65
interest groups, 27
interface, 40, 41, 99
international standards, 20
Internet, x, 11, 59, 60, 64, 68, 69, 72, 76
interoperability, 8, 18, 25, 26, 50

K

Kentucky, 99

L

labor force, 68
lakes, 10
Land Cover Analysis Tool, 95
landscape, 7, 60, 75, 98
languages, 69
lasers, 79
LCAT, 96, 97
leadership, 2, 4, 28, 40, 49, 51, 52
legislation, 5, 25, 32, 47, 48, 50
lifetime, 86
local government, xi, 3, 4, 6, 8, 9, 18, 23, 25, 27, 30, 32, 41, 42, 44, 45, 50, 53, 54, 55, 56, 65, 67, 73, 76, 80, 81, 82, 93
location information, 62
logistics, 72

M

majority, 53
management, x, 2, 5, 6, 7, 17, 18, 19, 22, 25, 27, 39, 42, 43, 44, 46, 47, 49, 59, 72, 87
mapping, ix, x, 2, 8, 9, 21, 23, 26, 36, 53, 59, 60, 64, 65, 68, 73, 76, 92, 94, 99, 100
MapQuest, 60, 64
marketplace, 27, 42, 80, 82
media, 94
membership, 2, 4, 33

mercury, 94, 95, 97, 100, 101
meter, 31, 38, 68, 77, 78
Mexico, 92
Miami, 98
Microsoft, 67, 75, 77, 78, 79, 82, 86
migration, 66
military, 13, 62, 65
missions, 49, 52, 61
mobile device, 70
modeling, x, 59, 76, 99, 100
modern society, 72
modernization, 36
monitoring, 43, 44, 47
Montana, 85

N

National Aeronautics and Space Administration, 20
National Geospatial Advisory Committee, 27, 40, 42, 43, 49
National Geospatial Technical Operations Center, x, 87
National Park Service, 22, 24, 98
National Research Council, 9, 23, 28, 35, 36, 37, 40, 41, 56, 57
National Science Foundation, 20
national security, 5, 18, 33, 80
natural disasters, 43, 44
natural gas, 47
natural hazards, ix, x, 2, 17, 100
natural resources, 43, 69
navigation system, 3, 7, 60, 62, 64, 72
navigational assistance, 62
New Zealand, 100
next generation, 63, 88
NGAC, 27, 28, 29, 32, 35, 36, 37, 38, 40, 42, 43, 49, 51, 52, 56, 57, 83, 84, 86
NGOs, 27
NGTOC, 87, 88, 89, 90, 91
Nile, 72
noise, 13
North America, 92, 99
NRC, 23, 35, 36, 37, 40, 41, 42, 44, 48, 51, 52, 53, 54, 55, 56, 57, 58, 86

O

Office of Management and Budget, 2, 4, 20, 35, 40
online information, 33
operating system, 67
opportunities, 44
orbit, 7
outreach, 92
overlay, 96
oversight, 34, 49, 53
ownership, 6, 10, 24, 26, 28, 42, 43, 44, 46, 47, 56, 73, 81

P

paradigm shift, 3, 75
parity, 32
permission, v, 62
permit, 47
personal computers, 67, 68, 72
photographs, 7, 21, 67, 95, 96
pipelines, 18
platform, 76
police, 7, 72, 100
policy choice, 100
policy initiative, 29
policy makers, ix, 1, 11, 17, 99
policy options, 51, 53, 94
policy problems, 83
political leaders, 55
portfolio, 99
price changes, 98
primary data, 9, 27, 80
private firms, 55
probability, 99
producers, 2, 3, 18, 30, 47, 53, 54, 55
productivity, 27, 80
profit, 22
project, 24, 34, 43, 45, 46, 98
prototype, 24, 98
public access, 88
public domain, 27, 31, 51, 53, 54, 55, 80, 82
public sector, 75
public service, 8

public-private partnerships, 80

Q

quality of life, 98
quality standards, 24
query, 72

R

reactions, 62
real estate, x, 13, 39, 56, 74, 82, 98
real time, ix, 1, 11, 75
recognition, 75, 83
recommendations, v, 5, 27, 28, 29, 37, 38, 41, 43, 52, 54, 57, 58, 83
redundancy, 30, 40, 48, 51
remediation, 46
remote sensing, 7, 8, 80
research and development, 88
resolution, 8, 9, 21, 24, 31, 68, 74, 77, 78, 82
resource management, 51
resources, xi, 21, 25, 26, 38, 40, 43, 47, 50, 52, 77, 86, 93
runoff, 6, 10

S

satellites, x, 7, 8, 21, 59, 61
sea level, 79, 83
Secretary of Commerce, 29, 51
Senate, 25, 48, 50
sensation, 60
sensing, 8, 80
sensors, 64
servers, 69
sewage, 43
shape, 62
signals, 7, 61
slavery, 62
social network, 64
software, 65, 66, 67, 68, 69, 75, 97
space policy, 80
spatial information, 2, 3, 18, 76
species, 95, 97
specifications, 48

spreadsheets, 69
stakeholders, 17, 22, 26, 27, 31, 46, 57, 75, 83, 98, 99
State Department, 36
statistics, 17, 97
stimulus, 33
storage, 7, 47, 67, 95
streams, 10, 48
suppression, 46
survey, 7, 8, 46, 56, 62
sustainability, 46

T

taxation, 8
telecommunications, 18, 43
tissue, 95, 101
total costs, 6
trade-off, 75
training, 37, 75, 88, 89
transformation, 7, 74
transparency, 69
transportation, 19, 21, 47, 91, 95

U

U.S. economy, 32
U.S. Geological Survey, vii, xi, 19, 35, 36, 45, 87, 93, 94, 98, 100, 101
universe, 6
universities, 22
updating, 8, 81
urban area, 31

V

vehicles, 5, 62
vertical integration, 61
vision, 19, 30, 40, 41, 63
visualization, 3, 16, 75, 76, 79
vulnerability, 17, 99

W

walking, 65
waste, 30
water quality, 95, 98

watershed, 6, 67, 96
wealth, 91, 99
West Virginia, 47
wetlands, 10

wildfire, 46, 47
wildland-urban interface, 46
wildlife, 94, 95